The Levels

All You Need to Know to Master the Art of Poker...

By Yadi Javadi

SETTING THE SCENE

This *could be* the greatest strategy book of all time.

Primarily it serves as a step by step study guide for all those who want to know how to play Poker.

If you're new to the game, this book will guide you through the fundamental theory which will enable you to contend with any opponent. If you're a seasoned professional, here you will find all the defining factors involved in an exploitative poker decision.

This book teaches you how to think through poker plays.

It teaches you how *all* players think through their poker plays.

Knowing how your opponent thinks through their plays is ludicrously beneficial to a poker player, especially if the opponent doesn't know it themselves. This will give you a *huge advantage* at the poker tables, it will make you loads of money, your confidence will soar and yet still, it's not the best thing about this book.

The most beneficial thing any strategist can learn is how to use this thought-process *for themselves*.

This strategic method has been evolving for billions of years to become perfect in almost every way. It is incomparably efficient. There is no easier way to concoct any conceivable strategy.

If that doesn't sound remarkable enough, you will also find this logic extremely easy to learn and just as easy to apply. Meticulous care has been afforded to the design of this book to ensure that every piece of information is delivered to you in the correct order at the correct time to minimise your effort and maximise your gains.

This is the holy grail.

And so it's time to say bye-bye to poker theorists like David Sklansky. Times have changed.

I give to you the true secret to mastering the game of Poker, and indeed any game of life.

Yes, I did just say that.

In this book I describe *the* strategic method used by human beings. I outline *all* the factors involved in making a strategic decision.

Why can you trust me?

Well, I was heavily involved in the game of Poker as a player and a winner and that was before I realised why I was so successful.

Like all decent folk I am now possessed by the aching desire to share my knowledge and skills with others so that they too can master their own game.

No-Limit Holdem (two cards for each player) is among the simplest forms of Poker, it's also the most popular version of the game and has always been my game of choice; hence, I use it here to illustrate the theories and concepts of *The Levels*.

Enjoy, embrace, expand and prosper…

TABLE OF CONTENTS

SETTING THE SCENE i
   THE INNOCENT GAME 1
   THE PROFESSIONAL POKER PLAYER 4
   THE MASTER 12
   INTRODUCING THE LEVELS 16
   DEFINING A GAME 20
   COGNITIVE DISSONANCE 22
   ADVANCING THROUGH THE LEVELS 25
   TIME 29

LEVEL ONE 31
   OUR CARDS 36
   THE BOARD 40
   OUR POSITION 44
   OUR CHIPS 46
   OUR BODY/EMOTIONS 49

LEVEL TWO 54
   THE VILLAIN'S RANGE 58
   HIS BOARD 63
   ADDING THE CHIPS 67
   IMPLIED PROFITS 71
   THE EASIEST WAY 73
   THE VILLAIN'S BODY/EMOTIONS 76

LEVEL THREE 79
   OUR PERCEIVED RANGE 86
   BLUFFING 91

PRE-ADJUSTING	94
BET-SIZING	100
THE STANDARD PLAY MODEL	104
OUR PERCEIVED BODY/EMOTIONS	112
LEVEL FOUR	**114**
LEVEL FIVE	**121**
THE FAMOUS CONTINUATION BET	126
LEVEL INFINITY	**130**
ONE LAST THEORY	139
ABOUT THE AUTHOR	145

THE INNOCENT GAME

The famous Poker theorist David Sklansky started his book by describing how *beautiful* Poker is, he wrote that; "On the surface it is a game of utter simplicity, yet beneath the surface it is profound, rich and full of subtlety".
Over the last 30 years his book has sold millions of copies, his theories have been adopted by an entire generation of poker players and his name has been written into legend.
I myself once thought his work was excellent, but to my own surprise, I've come to realise the old legend had absolutely no idea what he was talking about.
His fundamental strategy book was fundamentally wrong.
Poker is *not* some beautiful game that fell from the heavens. And nor was it forged in the fires of hell. If I had to come up with an analogy I would say that Poker came from the mud.
It is *clearly* a dirty game. I mean seriously, it's a gambling game and you make money by tricking other people into giving you their cash! This game is diabolical.
But it isn't all bad. Poker truly is a game of skill. It really is very similar to a sport.
The players don't use many of their physical attributes, but we do use all of our mental attributes - Bravery, confidence, accuracy, creativity, moral. All of these are important to a poker player and will be developed whenever you play.
Brain training is great, but the thing I like most about this game is that you don't need to be Einstein to play it amazingly well.
You don't need to be a mathematical genius to make a killing at the tables. And nor do you need to read facial expressions like some kind of *Marvel hero*.
All you really need is to have a good head on your shoulders.
And that, right there, is the key to Sklansky's biggest mistake.

That "beauty" he wrote about didn't belong to Poker. That subtle, rich and ingenious thing he described, that wasn't Poker...
… It was the *mind!*
In actual fact – Poker is nothing but a ludicrously simple card game played using the mind.
It provides us with a platform on which our minds can compete.

The only things that *really* belong to Poker are its *rules*
The same is true of all games.
Learning a games rules is usually very easy and takes only a few moments. Once learnt, we *will* feel as though we know how to play the game, but, there *are* many things the rules don't tell us.
They don't tell us when to bet nor how much to bet. They won't tell us when we need to fold.
Rules don't teach us how to strategize.
That is a job for the mind.
With our amazing minds we humans are able to strategize in any type of game *without even noticing that we're making any kind of calculation.*
Making strategic decisions, like betting, often seems effortless for us. It *seems* as-though there's nothing to it. But deep in the dark recesses of our mind there is a very specific process taking place. When we memorise a game's rules our subconscious mind secretly assimilates the conditions into a formula used by the mind to make strategic decisions.
It's this natural, built in, universal strategic *formula* which enables we humans to strategize.
And so, if we want to know exactly how to play a game, we need to understand not only it's rules but also this diaphanous formula.

Understanding this formula could be the key to understanding everything. But, unfortunately, the mind won't give up its secrets without a fight
Your mind knows all your strengths and understands all your weaknesses. It knows if you're sensitive to vanity, anger or fear

and it often uses these weaknesses to convince you to avoid it's secret formula.

Luckily, if you're not too familiar with Poker you should feel little or no side-effects whilst considering the logic here. If I have done my job well you should find the concepts in this book insightful and will walk away knowing exactly how everyone thinks through their poker plays.

For the professional poker players it will be different.

The second an experienced player hears about this book their own mind will become their very own perfect enemy.

They will do anything and everything to discredit this logic in a desperate attempt to keep *themselves* from accepting this simple formula.

Some of them will say that the formula is obvious and that everyone already understands it.

Others will say it's way too complicated and is clearly nonsense.

Some will discredit the logic by attacking me with sly pokes.

Some might well attack you if you ever dare mention that you understand this stuff.

Dealing with these people is never a pleasant experience, and nor is it all that easy either. They are often formidable opponents.

Professional poker players usually have highly developed minds which they've trained for many, many years.

To play Poker is to submit yourself to an assault of emotions as win-streaks pound you to-and-fro

This toughens you up.

Your mind becomes a rock.

It's important to make sure that you don't leave your foot under that rock, for, soon enough, a challenger will step onto the mud in front of you.

THE PROFESSIONAL POKER PLAYER

Professional poker players have been subject to many stereotypes over the years.
First they were portrayed as cowboys in name and nature, tricksters who risked their life on every turn of the card.
Next came the casino high-roller, spending his days lounging around Vegas in Armani dinner jackets.
And more recently we have seen the rise of the intellectuals.

Many characteristics of the cowboy are prized by today's professionals yet there are few who would risk their entire bankroll in one night let alone one hand
Die-hard cowboys do still exist and always will, but they're not the professionals.
They're usually untrained players who support their play with an alternative income source.
They often converge in local casinos or in online tournaments where the skill level is low.
Although capable of true greatness, these players gain very little experience which usually leaves them outmatched.

Professional casino players are still around but they're sparser than ever
In most countries there are only a handful of these guys to be found.
Making a healthy living at the casino requires lots of money and so these guys do often drive suave cars and live the easy life.
They do often spend their days being waited on by Swedish models in stylish hotels.
They even develop the stone-cold faces and callous confidence you see in the movies.

These "James Bond" style players never really went anywhere, but they *were* overtaken.

With the birth of the internet everybody became capable of playing Poker at any time of day from the comfort of their home
Professionals realised that they could play multiple tables which multiplied their income.
They purchased software which improved their play.
Teachers and coaches learned to stream live lessons.
Smaller bankrolls reaped higher rewards.
The sheer number of people with access to the internet allowed for much larger tournament prize pools.
All of a sudden the internet became the preferable place to both learn and play this game.
The ability to read expressions became a minor factor. The professionals learnt to read statistics instead.
Swedish massages were replaced by sore backs.
With inflated winnings and artificial assistance, becoming a professional had become a lot easier. All that's needed nowadays is a basic computer, an interest in games and some small ability with numbers.

Internet players have many advantages over casino players and it wasn't long before they began outmatching their casino counterparts
Casino players will only play around 30 hands an hour, but an internet player usually plays around 400.
At 40 hours per week, a casino player will see around 60k hands a year while an internet player usually sees close to a million. All of which are assessed, analysed and recorded within their software.
The additional experience gained by playing on the internet is very valuable, but the software helps a lot too.
Tracking software feeds all the information it gathers on each

player back to the user in the form of statistics on a live HUD to aid the player whilst they sit at the tables.

Many other types of program are available too. Most of them act as training tools, some are priceless, others are worth less than nothing.

The best way to keep an eye on the ever-changing world of Poker is by visiting online poker forums.

Lots of information can be learned on competitive forums.

When players with different skills come together to discuss different theories the pool of information generated can be invaluable.

Being able to consistently make money playing a game is the defining factor which converts a player into a professional

Every professional poker player nowadays knows roughly which games they can beat and how much money they will make per hour on average at their tables.

Their profit and loss estimates are usually remarkably accurate which makes it reasonably easy for them to enter into staking agreements, which many do.

Lots of pro players hop from one casino to another chasing down the easiest opponents. After selecting a casino they'll then hop from table to table as they search out the exact seat which will be most profitable for them.

These guys know that the weaker players give them more money. They know that they'll make a better hourly wage if they're sat with less skilled opponents and so they search them out.

Other pro's will find one good seat at one good casino and spend the rest of their life taking money from the same opponents.

Others will play at the most difficult web-sites in a proud attempt to gain recognition.

All professionals know that certain types of game yield a more regular income than others

Ring games offer a more stable income, while tournament

winnings are more irregular.
Omaha is irregular, Holdem steadier.
Information like this is very important to a professional player because these people are just like everyone else.
They need to pay their bills!
Players who can only beat the smallest stakes will often find that it's a constant struggle to stay ahead of their rent. These guys will often grind hard for long shifts over many days.
Once capable of winning in medium stakes games, players become fully aware of the freedom this profession allows. With only a laptop and an internet connection they can make a very healthy living with very little effort. You'll find these guys sat in cafes on beach fronts all over the world.
High stakes games are hard to find. Which means that the high stakes players are usually restricted to the biggest websites or biggest casinos. These players often travel a lot, following the biggest competitions.
Not all professional players have irregular lifestyles, many have families and use poker to provide a regular income. Even more use their profits as a part time income whilst they work in other professions or study at university.

Professional poker players are not all geniuses
It's a misconception to think you need super human powers to play Poker well. Anyone can learn the game and win.
In fact, "ordinary" people even get picked up by poker scouts (there aren't many of these around, but they do exist and they do go around *headhunting* for the best players to win for them).
A poker scout isn't specifically looking for mathematicians or academics, they look for level headed individuals, down to earth people, brave and/or open-minded people.
Remember that it's our mental attributes that we'll be using to both learn and play this game. Things like bravery, confidence, moral and stamina all come into play.
Math ability does help a little, but there are very few sums and so

you will pick them up very quickly.

If you're currently working for minimum wage at a corporation that you care little about, I certainly would *not* advise you to quit and put all your money on the poker tables
My advice for you is to put a *tiny* amount of money on the cheapest and easiest tables and get your head around the logic in this book.
As you become proficient, you'll start to win.
You'll then have the money to move up to higher and more difficult games.
Slowly but surely, you'll build up money. You'll build a stack.
Once you start winning regularly you can choose whether or not to enter into a staking agreement. From this point onwards, you won't need any money at all.
Staking companies will pick you up and take all your risk away but they do also take a cut of your profits.
Using staking companies can often be worthwhile if you can find a deal to suit you. They often offer coaching programs which can be useful. They *will* usually force you to play a certain number of games/hands per week too, which can be annoying.
I started my professional career on a staking agreement and within the first year I'd bought my own stack. I personally found it was far more comfortable to use my own money. There was less pressure, not more. And I got to take home *all* my profit too. But each to their own.
If you're able to apply half of the information in this book you'll never need to work any job you didn't want ever again. You'll be able to make a career from Poker without even having any money in your bank.
If you fully understand everything in this book, you can do whatever you like.

Poker is a game played using the mind. If you'd like to develop or make money from your mind, then Poker is a great way to

do it
At the poker tables, you don't have to wait for your boss to give you the promotion you deserve.
If you're good, then you win.
If you're better, then you win more.
Poker is not a strict task master, you can work or study the game whenever you like.
It's for everyone and anyone.
From time to time I'm teaching my young daughter how to play. When she's 18 I expect that she'll be able to make as much money as she'll ever need by sitting in front of her computer whenever she likes.
I don't know why everyone's not doing it.
…Well... I suppose I do...

There is always a chance that we will lose
This isn't much of a worry for a professional player, but for a beginner the uncertainty can be daunting. And for good reason too.
Some *unprofessional* players have lost their family home playing Poker.
Some have ruined their entire lives.
Some people have even died!
The Pro's don't have some magic trick which grants them immunity to losing in this luck based game.
They do lose, and often too. They just know that the money always comes back.
They calculate their entire bankroll ensuring that they always have enough money behind to cover their losses.
If some disaster does befall them, they simply move down to cheaper games and go back to building a stack.
Losing does still sting them, which can make this a rather uncomfortable profession. Over time the downs don't affect a person so much but losing money still hurts the best of them at times (even though they won't admit it!).

The professional doesn't expect to win every day, many don't expect to win every month even.

The very best player in the world will know that there's a chance he could make no profit after tens of thousands of hands.

This isn't a problem for him.

Pro's consciously aim *not to be* results-orientated, they try to concentrate on the correct factors *at all times*.

Whether they're winning or losing, rejoicing or crying, they know they must keep their mind on the cards.

Ultra-high stakes players seen on TV do obviously exist but I deliberately haven't spoken about them

Once you find yourself famous your career changes dramatically. The TV player will not be searching out better tables and planning their hourly earnings. Instead, he or she might spend their time preparing themselves for some interview or article.

I'm not taking anything away from these players. Some of them look like they might well be better than me at the tables.

It's just that their career is somewhat alien to that of the "regulars."

Every *Reg* will dream of reaching a final table in a major event and thus gaining fame, fortune and a sponsorship deal.

But even the best of them know that they have little chance in such large, rare and expensive competitions.

To enter these £10,000 per ticket tournaments, most Regs wouldn't feel safe unless they were sat on at least £500k.

Most Regs won't ever be able to afford those games.

Professional players sit at the tables they can afford to and play against a group of players that they can usually beat

Sometimes they push for higher, more expensive and more difficult tables.

Regs aren't all daredevils and most have never been to Vegas. They're not all chasing massive jackpots.

The vast majority will spend their days sat at their computer as they grind away at their profession so as to earn the money they require to live.

Someday, after a few years, when they have enough capital behind them to enter major events, they are sure to start taking shots at the biggest tournaments.

And, sooner or later, they know they'll win.

THE MASTER

In this book I deliberately teach as few strategies and plays as possible. I concentrate only on the theory behind the plays.

The perfect master would be able to comprehend all strategies, all plays, all opponents *and* they would understand all the theory.

I can teach you how to play, I can teach you the theory, but to contend with the best at the Poker tables all of your mental attributes will come into play.

If you have no empathy you will struggle to understand the individual opponents. If you're not creative, you will struggle to concoct abstract plays.

Remember that Poker is just a game that uses the mind. The better the mind, the better the player.

To master Poker is to master the mind
Throughout history there has been an endless number of individuals who have developed our understanding of the mind and its different abilities.

The first name that jumps out at me is the famous Greek philosopher Socrates. He once said:

"The only thing we can be certain of is that we cannot be certain of anything."

Plato studied alongside this great man and has highlighted many truths about the mental processes that we're all subject to.

Many a military general has helped us develop our understanding of the mind by showing us our strategic capabilities.

Countless artists have tuned into the mind to toy with our perception of everything.

I'd bet that all the *greats* would be amazing at the poker tables. All of them seem to have gotten close to mastering the mind.

Mastering the mind appears to be an impossible task, however, there *is one guy* out there who might well have pulled it off... Somebody I haven't mentioned yet...

2500 years ago there was a man called Siddhartha Gautama who understood the mind and its capabilities unbelievably clearly
This man claimed to have completely unlocked his own mind, and in doing so, claimed to have gained access to all knowledge.
Quite a bold statement wouldn't you say?
Surely it couldn't be true... Right?
This is from Einstein –
"According to general relativity, the concept of space detached from any physical content does not exist."
And here is Siddhartha describing the same thing –
"If there is only empty space, with no suns nor planets in it, then space loses its substantiality."
Huh! How could he know that?
There are hundreds of these unbelievable quotes from this dude. Here's another –
"All such notions as causation, succession, atoms, primary elements...are all figments of the imagination and manifestations of the mind."
How the hell did this guy know what an atom was!?
He lived 500 years before Jesus was born yet he clearly understood specific facts regarding quantum physics. How could that be?
It doesn't stop there...
He described the development of a baby whilst inside the womb. He explained the expansion of the universe. And he said thousands of things about the mind.
This dude didn't have a laser guided microscope, and nor did he perform any crude experiments on pregnant women. He says that the way he learnt all these ridiculous things was by sitting under a tree and meditating.
Once his mind was perfectly calm and therefore free from all emotional interference, he claims to have unlocked his true

cognitive capabilities which go way beyond anything that we've learned through science.

He also said that each and every one of us is capable of acquiring this same mental state.

He then spent his life explaining how we'd go about it.

He explained how to develop concentration.

This man truly is the master

He's been revered by millions upon millions of people since he lived.

There are countless statues depicting his image.

Everybody's heard of him.

When he acquired *the perfect mind* he left his old name behind.

He said that he was no longer Siddhartha.

He said that he had become a Buddha.

Scientists have always revered his teachings, but most of us *common folk* disregard his lessons before even learning what he had said

This is partially due to our own arrogance, but I place most of the blame on the way religions have been used as a method of control over thousands of years, killing and maiming millions of people while demanding blind obedience to logic which doesn't often make sense.

Buddhism is not like other religions.

Buddhism *makes sense* of the other religions.

The Buddhists don't say things like, *"abstain from lusty thoughts or you will burn in hell for all eternity!"* they say that if we do lust after things we will find it more difficult to concentrate. Our mind would be clouded. This clouded mind could then lead us down dark paths.

The primary reason for all Buddha's lessons was to help us develop concentration so that we too could reach the same state as he.

His lessons were often surreal and quite hard to comprehend, but once deciphered it's often a surprise to find that they do seem to make a great deal of sense.

I definitely don't fully understand what he was trying to explain
But I do see the value in developing our concentration. For Poker players it is huge. If we can concentrate on the correct factors while we're at the tables we can't go far wrong.
If we *cannot* concentrate on the correct factors while we're sat at the poker tables, we can quite often ruin our entire poker career.
In this book I will teach you those factors. I'll teach you all the things that you need to think about while making a Poker decision. But if you can't concentrate in general you will still struggle to play.
Meditation is an exercise Buddha said is certain to improve our ability to concentrate and it is quite easy to do – you simply centre your awareness on any one thing, and then when your mind wanders you bring it back. Meditation. Simple.
You train yourself to stay on topic. You practice concentrating.
Playing Poker also helps us develop concentration. Yoga seems to help a lot with this too. As does keeping fit. Or being nice. There are all kinds of ways to improve our ability to concentrate.

If you want to understand how to develop any of your mental attributes Buddha's lessons are the place to go
In this book all that I do is explain how to use the mind to play a very simple game. He described how to use the mind in life. He described the mind in general.
I'm still delving into his stuff myself but from what I see so far, his lessons seem to match up to the logic I describe scarily well.

INTRODUCING THE LEVELS

All strategic decisions are broken down using the Levels.
When a boxer fakes with his left and follows with his right, he's using the Levels.
When a petit tennis player screams as she blasts the ball to the back of the court, she's using the Levels.
When a shopkeeper raises the price on his most popular product, he's using the Levels.
When the Chinese built their wall, they did so in accordance with those same Levels.
The Levels are followed by parents as they punish their children.
They're embraced by the nurse who stays at her patient's bedside.
They are that which leads to lions having claws.
The Levels don't belong to any one particular game; they belong to the mind.
They make up the formula that the mind uses to calculate decisions.

Professional poker players are making tens of thousands of strategic decisions every day
I calculate that while playing online I personally make one strategic decision every two seconds or so.
Under these conditions it's not difficult to see the importance in understanding the decision-making process itself.
While others around me were busy memorizing more and more complex plays, I was sat in the sun considering the simplest of situations so that I could figure out the generic logic behind all plays.
I began mapping out the decision-making process for Poker, and soon enough, all I could see was the Levels.

THE LEVELS

The concept of these Levels in relation to Poker is well known
They were named the *Levels of Thought* and are acknowledged as a fundamental component of a style called *exploitative*.
Many years ago, while grinding away at the poker tables, I found a strange pattern within the *Levels of Thought*. I noticed simple logic which proved there was a specific order that we should consider each Level.
This meant there was a specific order for us to consider all the different factors for every individual poker decision.
It meant that there was a specific way to think through a poker decision.
All that was left for me to do was match up each individual poker factor to its proper Level and then I would have the formula for creating poker plays.
It looked so easy...

I had absolutely no idea what I was getting myself into
Finding that pattern in the *Levels of Thought* was like finding the first numbers on a Sudoku board.
One thing led to another, and then crashed into another, and then before I knew it I found myself gazing at the boundaries surrounding all comprehendible strategies.
All I was trying to do was find the perfect way to think through a Poker decision!
I wasn't trying to find the way to think through *all* decisions.
This came as a massive surprise to me!
I really don't mind admitting that I am 'under-qualified' to write a book on such a broad topic as this.
I am just your everyday bloke off the street who happens to have been open-minded enough to embrace new ideas and concepts while working away at my profession. Poker.
I am not much of an academic, I'm not a scientist, a psychologist or a mathematician; I'm just a Reg with an abundance of curiosity.

Having stumbled upon way more than I'd bargained for, I set about validating my theories with my peers

And that was not easy.

I found it ridiculously hard to get any coherent contributions from any other poker players due to the sensitivity of this content. The vast majority of players violently rejected my logic. They didn't help me like I had hoped they might, instead they have hindered me at every step.

I'm sure there are some players out there who could have aided me in deciphering these Levels but as I write this now there are only a handful of people on the planet who are aware of how to play at Level Four. If I could have spent a few months with a few of these fellows then I'm sure my work would have taken a fraction of the time that it did, but these fabled players are hard to find at the best of times. I had to work on this by myself.

Eventually I found a great editor who knew nothing of Poker and with his help I created this book.

This book was completed independently and my budget was very small

But who cares! I've believe that I've been successful in mapping out the minds own decision-making process which I have matched up to Poker.

I've tested it countless times, it has stood up against all kinds of assaults and now I am ready for the world to see it.

I am ready to introduce you to *The Levels of the Mind*.

- **Level One** – contains our real, physical presence within the game

- **Level Two** – relates to our perception of our opponent, the "Villain", within the game

- **Level Three** – is our perception of our opponents understanding of us, within the game

- **Level Four** – is where it starts to sound *real* confusing.

- **Level Five** – sounds ridiculously complicated but it's still pretty simple for we human beings.

- **Level Infinity** – is not like the others; it's commonly known as G.T.O. Game Theory Optimal. The unbeatable strategy.

Each subsequent Level contains all the factors from all the previous Levels.
It grows like a pyramid being built from the top down.
The later the Level, the more information there is to consider for it also contains all the logic from those before it.
But don't let that put you off!
Once you start to take it all in you will soon understand each one quite easily. It really isn't rocket science.
I promise that it will all come naturally to you.
Believe it or not, you can already apply all these Levels like a master.

DEFINING A GAME

A game is defined as –
"A competitive sport or activity played according to rules."
There are only two requirements for any activity to be a game.
It must be competitive, which means that it must have some kind of goal.
It must also have rules.
That's all.
The common understanding of a game is an activity or sport which has rules that have been deliberately made for that specific game by some person.
Games like Football, Monopoly, or jigsaw puzzles perhaps, all have specific rules and goals concocted by a person.

The *mind's formula* does not exclusively work with made-up kind of games, it caters for *every situation* which has rules and a goal
When we start to consider rules like gravity or "Thou shalt not steal", it's clear that every situation we might find ourselves in has rules. Throughout our day to day lives we are quietly bound by many such rules which are all stacked up on top of each other in our minds.
Exactly the same is true of goals. If we had no goal then we wouldn't get out of bed. We wouldn't have gone to bed in the first place.
Every action we make works in-line with both goals and rules whether we notice them or not.
All decisions are made using them both, but usually we're not aware of the fundamentality of their existence in our minds.

Rules and goals hold a key place in the minds formula, they

are vital to all strategic decisions
But maybe they are even more than that.
These two factors are at the forefront of all our happiness and all our sadness. They seem to shape our understanding of everything. They are everywhere. And so, everything can be considered as a game.
If anyone ever tells you that life isn't a game you're free to explain all this to them if you like, but, that's exactly the kind of information that many minds will reject.

COGNITIVE DISSONANCE

We humans habitually meander through life making all kinds of strategic decisions without giving any thought to the thought processes that allow us to achieve these truly amazing feats.
We're perfectly happy to leave all the work to our ever-loyal subconscious mind.

Throughout our day to day lives the subconscious mind uses the Levels to strategize for us
We innocently throw in information and the subconscious carefully stores away the knowledge in its proper place.
We feed in problems and the subconscious uses the Levels to show us our different solutions.
The subconscious usually takes care of this stuff *for* us, to make our lives easier.
If we do decide at some point that we want to consider our strategic thought process, *as standard,* the subconscious instantly grants us conscious access to whichever part of the formula we like.
Consciously recognising factors related to the Levels, as standard, is not a problem.

The problems in accessing this information only arises because we, consciously, have control of our own minds, and, in our ignorance, we make mistakes
Any time we learn how to play any *game* whilst unaware of the minds strategic method it's very likely that we will accidentally build trust in strategies that *do not* work in line with it.
We'll begin to trust in strategies that conflict with it.
This conflicting knowledge then initiates something called *cognitive dissonance*.

THE LEVELS

Cognitive dissonance is a reasonably well-known term which describes a negative feeling experienced by a person who holds multiple contrary views at the same time while confronted with information which conflicts with one of their beliefs.

It is also well known that the person would actively attempt to avoid situations or information likely to increase this negative feeling.

In short, when a person has learnt to trust in strategies or plays before understanding their own natural strategic thought-process, their own mind is likely to protect the knowledge by stopping the person from understanding their own strategic thought-process.

The more trust we place in conflicting strategies/plays for a game, the more aggressively the subconscious will covertly dissuade us from recognising its all-important formula in relation to that game by making us feel, in some way, bad

And that is exactly why experienced poker players will struggle so much with this book.

If we've been memorising Poker plays and building trust in that knowledge our mind will protect the knowledge by stopping us from seeing how we calculate Poker plays.

If we've been memorising ways to act at the dinner table, our mind will make it difficult for us to understand the reasons *why* we act in that way in that situation.

It is the real reason *why* that we lose by forsaking this formula.

If anyone then challenges our impure trust in any way, perhaps by showing us other 'plays', our mind will reject the information by using all our negative characteristics to distract us from seeing the true reasons for our own actions.

The mind distracts us from its formula by making us angry, confused or subtly scared.

Throughout our lives we are all subject to these kinds of feelings

Is it just a coincidence that not a single one of us consciously understands the *reason why* we're alive? None of us are aware of the reason why we are playing this game of life.

It seems clear to me that all our confusion comes from all the trust we've placed in all the knowledge we've gathered regarding all strategies throughout our entire lives.

And I think Buddha agrees with me.

I expect that if we *were* perfectly open-minded we *would* be able to understand the meaning of life easily. But we all went to school. We've all memorised all kinds of things. We trust in lessons from our parents and many peers. We've all learnt loads, and loads, and loads of strategies that may or may not work in line with the Levels.

If I understand Buddha correctly, he says that not all strategic knowledge is bad for us. He says that there is a kind of perfect knowledge, a type of knowledge which we can place our trust in that will not hinder our ability to strategize. He described it as 'right knowledge'.

He said that the second we consciously come to understand all of this right knowledge we will see how life works and then all our negative emotions will disappear and we will be granted a feeling of enlightenment.

ADVANCING THROUGH THE LEVELS

We develop our understanding of our game by becoming aware of the Levels one by one.

We *are* able to learn individual aspects of a Level before completely understanding the last, but in general, we advance through the Levels by breaking through our cognitive dissonance to unlock them in chronological order.

In the instant you come to recognise all the key aspects of a new Level you will suddenly feel that you now know your game.

In this moment of realisation you're granted a feeling that can only be described as enlightenment. From that moment onwards it appears as though a veil is lifted and your game time becomes far more pleasant.

It really does feel like you're a video game character who's just killed his hundredth zombie. You do seem to *Level up*. You feel stronger. You are stronger.

In one glorious moment you see everything about the game more clearly, and you win more too.

Unfortunately, every time we break through to a new Level, due to feeling that we now know the game, we will usually reinstate the trust in the knowledge which made the last Level advancement difficult

Beginners can sidestep this problem easily by pushing themselves, or coaches can push their students, through the Levels quickly without allowing time to settle at a specific Level until all have been consciously grasped.

In this way all the Levels can often be learned in a matter of minutes.

If you're a beginner, please do be aware that understanding

the Levels is not quite the same as being aware of each Level for every decision
For a little while after you grasp them in relation to your game you'll probably have to keep reminding yourself of each Level while actively playing.
But don't worry, provided you recognise their importance for *every* decision you'll notice them lodged in the back of your mind within no time.
Developing an unconscious grasp of these Levels is very, very easy once you come to consciously recognise them.
Remember, you already know all these Levels. All of us are proficient in their use. We use them all day and every day.
So, we don't really need to *learn* them.
We just need to *notice* them.

Advancing through Levels One and Infinity is slightly different to the others
You won't feel a wave of euphoria when you recognise Level Infinity, which does seem to make sense to me and I think it will to you too.
And Level One is learned in stages rather than all in one go. We unveil the different sections of Level One as we come to recognise the higher Levels.
The first time we feel we know Level One is when we first learn the rules to a game. At this stage of our development we *do still use* all the higher Levels, we just don't normally notice.

Without any conflicting knowledge related to Poker dragging them down it's clear to me that a completely untrained player is far more capable of making, say, a Level Three poker play, than an experienced player who only consciously recognises Level Two
The only reason beginners haven't been obliterating Level Two players is because the knowledge the experienced player gathers isn't all problematic or completely inaccurate.
As I write this now, most Level Two players will have already

learned almost every Level Three factor at least as well as I will describe them.

They will also have learned many, many strategies and plays.

It's not what they know that's the real problem here, it's what they don't know. They must be missing one key piece of theory from Level Three or else they would have cracked the Level.

The knowledge of strategies and plays they have gathered whilst only being consciously aware of Level Two is making it difficult for them to approach these missing pieces of theory.

Cognitive dissonance attacks.

The amount of trust in conflicting knowledge related to our game determines both the difficulty of each Level advancement as well as the intensity of the enlightened feeling

Beginners reading this book will find it easy to recognise the factors relating to any Level, but they'll only feel a slight wave of euphoria after fully cracking each one. It won't seem like a big deal.

Experienced professionals will usually struggle to approach any factor from a higher level than that which they are currently aware, but, the instant they come to accept a new level they are likely to feel elated.

The benefits are relative too; as a knowledgeable player advances he'll learn the real reason behind all the knowledge that he's previously memorised. He'll suddenly understand everything he knows better. Which often makes a huge difference to him.

Beginners won't really know a great deal to begin with and so instead of a massive and sudden improvement, when a beginner learns a new Level they'll have attained the correct foundations on which they can come to learn, or concoct, specific strategies or plays for their game.

If you're an experienced player who suspects you may not be the highest Level of player then fear not

In this book I've separated all the key pieces of poker theory, given them their own chapter and placed them in their proper

Level.

In a way it's very easy for you to pinpoint the exact chapter which your mind rejects, but if for whatever reason you can't find it, there is this other technique you can try:

A great way for an experienced player to break through to a higher Level is to write out an individual play, one individual decision, ensuring that all the information is organised correctly (as I describe later in the book in the chapter entitled "The Standard Play Model").

While organising your own play on the page you unwittingly organise your own thought process and then the Levels are magically revealed.

If you collect many of these perfectly written out plays, please do put them together into a playbook and send it through to me. I'll be happy to help you publish your work.

Now that the standard play model has been established all Regs can put together their own playbooks.

This is an excellent way for all of us to give something back to the Poker community whilst gaining recognition as well as a passive income.

TIME

Time, the element of change, provides the building blocks which make up the Levels of the mind.
It seems to me like there should be a Level Zero in which there is only time because the past, the present, and the future are represented in each section of each Level.
Time is everything here.
In a way, the only reason this formula exists is to help us make sense of time
To help us make sense of changes.

The past dictates our present situation, but if we considered it properly during it's time then we wouldn't need to consider it in the present.
The future is used to calculate improvements to our present situation. If it's considered carefully then the present is expected before it appears and so little consideration is needed regarding our present situation either.
If we're capable of contemplating the future perfectly we'd never have to calculate either the past nor present ever again.

The past, the present and the future are used in the same way for all decisions
We don't simply think, "I want cake."
Instead, by using the past we're able to recognise our present situation, "What was that feeling I just felt? Ah yes. I am hungry!".
Then we consider our future situation after either eating or not eating the cake.
We recognise that we wouldn't be hungry if we ate the cake, that our present situation would become more positive once we have

a bite, and so we delve in.

Some of us look further into the future and decide that if we eat it we will get floppy ankles or clogged arteries. We might then conclude that this future would not be worth the satisfaction gained from eating the delicacy and so might decide against that option.

We might decide to make a different play.

Day to day calculations like this are usually completed subconsciously

We can watch our thought process as it happens if we want, but we don't usually bother. We don't usually interfere with the strategic process itself during our decisions.

We don't watch the engine, we drive the car.

You see, the conscious mind is more than this formula alone. It's like an awareness, an awareness that wields an all-encompassing power.

This formula is like a filter that we use to make sense of the changes that happen to the six things that we're *aware* of:

Sights, sounds, feelings, smells, tastes and thoughts.

We make sense of everything using this formula so that we can see what we might do to make changes of our own.

There is one other *time* that has a place within every Level. It's called the future-present

The future-present becomes very important as the Levels progress. It relates to the instances when we'll have a chance to make a further decision in the future.

Most games operate in real time and so we make decisions constantly. Even in these games the future and the future present represent different factors.

Any change in circumstance possible in the future gives us reason to reconsider our condition and decision after the changes have materialised.

We consider our next move, and the move after that.

LEVEL ONE

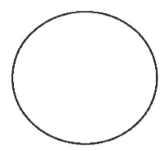

Level One contains our real, physical presence within the game

Level One is a sum of all the physical factors that belong to *us* in our game.
This Level can be understood very easily – if it belongs to us, is a part of the game and is a real thing, then it's a Level One factor.
Our chess pieces would be our Level One self in chess.
Our feet in football.
Our hands in boxing.
Our subconscious makes Level One *appear* to be very straightforward - we simply think about what is *ours* in the game and our subconscious shows us our Level One self.

At first glance Level One seems simple but once I started asking myself how and why these things belong to us I quickly realised that this is by far the most complex of all the Levels
To understand Level One perfectly we would need to understand

the boundaries of our own self.

What we truly are.

Ever since I was a teenager I've been trying to solve that riddle and it's a riddle that I'm still trying to solve today.

I definitely don't fully understand the *self*, but by understanding how it works in regard to Poker, I am starting to feel like I'm getting close to an answer…

(If this is the first time you're reading this book you may want to skip past the rest of this chapter. The next three pages will not help a poker player make decisions. The remainder of this introduction to Level One is full of speculation, it's also very philosophical and does not reflect the nature the remaining content. I toyed with the idea of putting this in at the end of the book rather than the beginning, but I resolved to give you this here warning instead. Please, be well aware that I certainly *do not* fully understand Level One. I feel no shame in saying that, for this is the Level that merges all 'games'. If you can understand it clearly, I suspect that you would understand everything).

Level One consists of all the things that we can use to help us win in our game

In some ways the entire universe is a Level One factor, as everything contributes in some way towards our winning.

The Sun, for example, is very important to every game on the planet. If we weren't using its light, we wouldn't find ourselves winning very many Poker hands.

We could say that the Sun is a part of this overall thing that we call *ourselves*, but the Sun is also a pretty irrelevant variable to consider during a poker decision.

At Level One of the decision-making formula our mind deduces the *relevant* factors from the 3D world which we can use to help us reach our goal.

It's the rules and goals that we use to deduce these *relevant physical attributes*.

When we acknowledge rules to a game like Poker, we're teaching

THE LEVELS

our subconscious which *restrictions* it needs to apply to its understanding of our normal self so as to comply with the individual game's specific conditions and goals.

Here are two examples to help you understand how this works –

1. In life we're aware that we can move anywhere on the planet. The whole planet is ours!

But when we follow the rules to Football, we learn that we can only move around the pitch. The only part of the 3D world which the mind deems relevant for this game is the grassy pitch.

Our mind simply restricts our physical capacity in coordination with the rules to the game.

2. As we go through our day to day lives, we're free to make decisions whenever we like, but in Poker the rules state that we must wait until it's our turn.

Again, we restrict our physical capabilities in coordination with the rules to the game.

Every time we learn a rule, whether we're learning a new job or how to deal with a certain person, our mind always *restricts* our understanding of *our self* so that we're able to cater for a more specific situation.

Poker is a very simple game and so when we come to deduce our relevant physical attributes we need not worry about much

The weather won't have an effect on our winnings.

We don't need to worry about our teammates, nor our colleagues.

The fitness of our pectoral muscles isn't important either.

The only physical attributes that we need to consider, the only things that contribute to us achieving our goal in this game, are a handful of cards, a stack of chips, and a seat.

These three things make up our Level One self in Poker.

And so, to learn Level One for Poker, all you need to learn is the qualities of these three attributes.

Once you have established an understanding of your physical attributes for any specific game, to complete this formula you

must then re-visit each attribute from a new perspective at each new Level, considering the attribute in the past, the present, the future and the future-present at each Level.

Cards, chips and a seat are the physical things that we use to play Poker
And its only because we *use* them that they become part of our relevant self in this game.
And *that* means that it's actually the *manoeuvrability* granted by the physical attribute which gives the physical thing a place in the formula.
Manoeuvrability is heavily tied into goals, desire, awareness, emotions and many other "invisible factors" that are very hard to pinpoint and so seem to come from a non-physical world, or spirit realm, if you prefer.
I don't fully understand how any of those things work, but I do get the general jist:
In a game like football we use all kinds of physical attributes which allow us many different options for manoeuvrability.
Provided we're on the pitch we can run, stop, jump, kick the ball, step over the ball, the list goes on and on.
In Poker, there are only three things that we can do with our self at any given time – we can bet, we can check, or we can fold.
When I considered what Level One would look like without the higher levels I realised that at that stage of our development we'd have no concept of these three actions.
From a Level One perspective, without at least a subconscious understanding of the higher Levels, our three options wouldn't exist.
It's only the potential for manoeuvrability that exists here at Level One.
And that got me thinking.
Perhaps it is only the potential for a physical self that exists at this Level too.
Perhaps our physical self doesn't *really* exist until we come to make sense of our capabilities at Level Two.

And that's not just an idle thought.

All we need to learn at Level One is the potential for manoeuvrability and the potential strength of our physical self.

We don't learn to actually *do* anything at this Level, we just learn what we *could* do.

Until we reach Level Two and start formulating plays there seems to be no substance to our physical self, there seems to be only possibility.

It's very difficult to be sure of exactly what Level One would look like without the higher Levels

The higher Levels are built into our subconscious and we have to use them whether we'd like to or not.

If the subconscious were not taking care of those higher levels for us we would have no concept of anything.

There are objects in the universe that observe things from a Level One perspective - Things like rocks, minerals, atoms and such.

A rock's reasoning is non-existent and it abides in the universe only as a physical presence within a set of rules.

Rocks exist only at Level One.

OUR CARDS

As we advance through the Levels we learn how to use attributes like our cards, but at Level One we need only learn how *strong* are cards are in relation to all that they could be.

The strength of Level One attributes is very important to every game. Without value, our attributes would never be better than the opponents and so we would never be able to win. Every encounter would end in a draw.

If some of our cards weren't stronger than others, or if being hungry held the same value as being satisfied, none of our deliberations would have any purpose, we could reach no goal, and so would have no need for this formula.

In this chapter, I'm not going to explain the basic poker hand standings for the sake of the readers who are not planning on playing this game. I won't explain which hands beat which, but if you need to know then please do look it up.

To teach you the range of strength of the cards, I need to explain how often the different hands will appear

In Holdem each player is given two cards.

We need to learn how often we will hold A, A.

How often we will hold 2, 3.

I won't be filling your mind with confusing sums and useless figures regarding the thousands of combinations of cards we could have.

Instead, I'll show you using percentages how often you will see each type of hand.

I find this is by far the easiest way to understand the probability of a hand arriving and it will also teach you to read a HUD on Poker-tracking software.

THE LEVELS

This is how often each hand appears in a full range –

Pocket Pairs appear 0.4% of the time each.

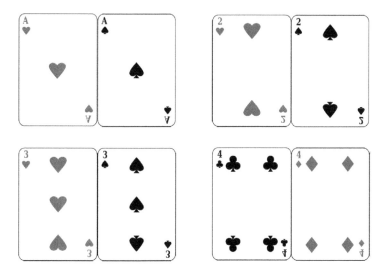

Non-Pocket Pairs appear 1.4% of the time each, and, from the Non-Pocket Pairs, Suited Cards appear 0.35% of the time each -

Don't worry too much about accuracy. I would simply remember that:

- Pocket pairs come up a little less than 0.5% of the time each
- non-pocket pairs come up a touch less than 1.5% of the time each
- and a quarter of the non-pocket pairs are suited

If we then come to guess what a range of cards containing the top 3% of hands might look like, we can do it reasonably easily -

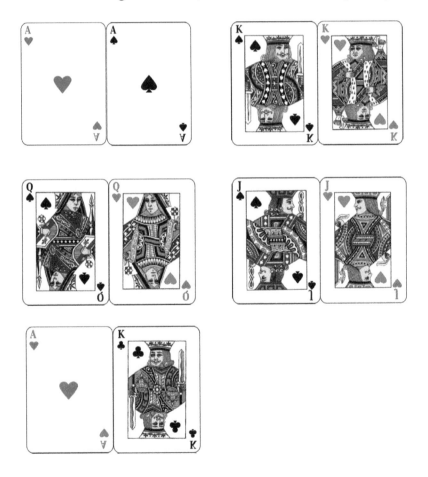

JJ+ = 1.6%

AK = 1.4%

Together, these hands add up to a 3% range.
When you come to play it is very important to understand roughly which hands the different size ranges will consist of
You need to be able to picture roughly what a 20% range might look like.
You need to be able to imagine roughly what a 50% range might look like.
Your maths doesn't need to be perfect, you just need a rough idea of what hands these ranges could consist of.
This is very easy to learn and will probably only take you a few minutes to grasp.
When we consider how future changes will affect the strength of our cards the sum does become a little more complicated, but not a great deal more so.

Step 1 - Learn the strength of our hand by learning where it stands in a full range of possible hands

THE BOARD

In Holdem, after receiving our two cards, and after everyone has had an opportunity to bet, extra cards are placed in the middle of the table for all players to use.
These extra cards are called *the board*.
The board is a shared attribute in Poker. All players can use these cards to strengthen their hand.
Shared attributes like this are common in all games.
I'm talking about the monopoly board or the football pitch. Perhaps the most important shared attribute in this day and age, in a general sense, is the customer.

To understand the true strength of our hand we need to know how strong it will be once the future has come about
We need to use a sum to guess the probability of a board coming that will make our hand stronger.

Let me start by explaining this sum at the end of a hand when the equation is at its most basic:
If we're on the fourth shared card, known as the Turn, we will only have one future card left to come (The river).
At this point we've seen six of the total seven cards that we can use to make our hand.
There are only 46 cards left in the deck and each has an equal chance of arriving.
We can easily surmise that there's a 1 in 46 chance of seeing each remaining card. We then use this information to calculate how often we would see a card which strengthens our hand. For example -

THE LEVELS

Our Cards =

Shared Cards (Turn) =

In this situation, there are 9 hearts left in the deck which would lead to us hitting a flush.
There are a total of 45 cards left, and so we have a 9 in 45 chance of hitting our heart.
45 divided by 9, gives us a 1 in 5 chance of hitting our flush.
In this situation, we also have a chance of hitting a straight.
This adds an extra 6 *outs* which would improve our hand.
Meaning that we now have a total of 15 from 46 cards which would improve us.
Around a 1 in 3 chance of hitting our card.
We might also consider how often we hit a single pair and add that to the equation in the same way.

Now, let's take the situation back a step and see how it looks with only three shared cards revealed. This stage is known as the *flop* **–**

Our Cards =

Shared Cards (Flop) =

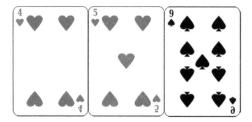

This time we have two cards still to come so we need to add a little to our calculation.

To keep it simple, we can just double the number of *outs* as we now have two chances to hit the cards.

I'd now say that we'd hit either the flush or a straight a little less than 2/3 of the time.

When we step back to the very start of the hand, back when we only have our original two cards (pre-flop), the same sum starts to become far too complicated to calculate in real-time.

This round of betting is known as pre-flop.

At this stage, there are still five shared cards to come which equates to around 250 million different possible futures.

There is no simple way to work out the future strength of our five-card hand when faced with such an equation. But at this stage we don't really need to calculate it anyway, and so I wouldn't worry about it much. Provided you understand how well your two-card hand stands up against the others you'll be fine.

If you do want to build up an understanding of how strong your preflop hand is likely to become, then by all means do look up different 'flop frequencies' just to give you an idea of how often the different boards will appear.

There is lots of software which can help you learn this stuff. Equilab is a good one that you can download for free to your PC, and there are also apps you can get for your phone. These programmes will show you the exact strength of your hand and its exact chance of improving to beat your opponent.

Step 2 - Learn how strong our hand will become in the future

OUR POSITION

One of the rules in Poker states that we're not allowed to take any action until it's our turn. This rule gives birth to a lesser attribute which is usually hidden to beginners –
Our seat, or our position.
In Poker, the last person to act during any round of betting holds an advantage. This advantage exists mostly because they are able to see what everyone else does before making their own decision. By gaining some idea of the strength of the opponent's hands the player *in position* is stronger, which enables them to play more hands more profitably than they could if they were sat in a less beneficial seat.

The most advantageous seat in Poker, the strongest seat, is known as the Button
If we're sat on the Button; during the pre-flop stage of the hand we're in a late position and on the flop, the turn and the river we're the very last person to act.
Imagine we're sat on the Button pre-flop, and then every player who acted before us has a weak hand and so folds.
When we look to our left we'd see that there are only two players left in the pot, both of whom will act before us on the next three streets of betting.
This already sounds good but it gets better.
The players with the worst seats are forced to put money in the pot before they even see their cards.
They have to put in the blinds.
These two players could still hold the worst cards possible and so the chips which they have already committed to the pot might well be free money for us to steal.

Having a better position helps us in other ways too
The most beneficial of these seems to be the additional control we have over the size of the pot.

If we're the last person to act then we get to choose whether to make a bet which will enable everybody to raise it up before we see the next card.

Or more importantly, we get to choose *not* to give everybody a chance to raise before seeing more shared cards.

If we're worried that a player who didn't bet might be planning to raise our bet, we can simply not bet.

Not only doesn't our opponent have a chance to raise us until the next street (flop, turn or river), he also misses out on one entire round of betting which can drastically reduce eventual the size of the pot.

When you come to explore your options after learning all the Levels you'll be able to recognise many different places where position offers an advantage
Position is a reasonably simple factor in poker, but it's also very important, even at the highest Levels.

Step 3 - Learn which seats are stronger than others. The stronger our seat, the weaker our hand needs to be.

OUR CHIPS

Chips are the most important attribute a poker player has in his possession.
Even if we're holding the best hand and sitting in the best seat, it won't help us much if we only have one chip left.
The strength, or the value of these chips is obvious. The more chips we have, the more valuable they are, and so, the stronger we are.
If we have lots of chips then we become more dangerous.
If we find ourselves with no chips the game is over.
Gaining chips is our primary goal for this game.

Every attribute exists to aid us in winning chips
Whichever cards we play and however we choose to play them, we do so for the chips.
If we did not see a potential to make chips then we'd give up on our hand. We'd fold.
Technically speaking the sole reason we make a play in this day and age would be to make cold hard cash. Which does make a slight difference while attempting to win a specific prize in a tournament, but in general, chips represent the prize that we stand to win or lose.
In Poker, the chips are the only attribute which can be transferred between players. We can win them from our opponent or we can lose them to our opponent.
As Level One is all about ourselves without the consideration of an opponent, there is very little to say about our chips in this section. Most of the stuff related to them is only unveiled at the higher Levels.

The Level One rules in relation to chips are simple:

- The chips allow us to take one of three options – we can bet/raise, check/call, or fold.

- When a hand reaches its conclusion, the strongest hand takes the chips.

At any given time we won't have more than three main options available. However, to understand our potential capabilities in their entirety we also need to take into consideration future opportunities to make plays.

We then find that there are potentially loads of different *lines* that we might take. There are loads of combinations of moves we can make.

As we come to understand the higher levels we will find the reason to take these *lines* but for now we simply need to understand that we can manoeuvre in Poker by making three types of play.

We can attack with a bet/raise, defend with a fold, or make the neutral check/call.

At the end of each hand, as well as the victor taking the winnings, the casino takes a share of the pot in the form of the rake

This rake is *usually* a percentage of the pot but the casino might take an hourly fee from each player instead. In tournaments the casino usually takes an initial entry fee from the players which doesn't interfere with their chip stacks.

Regardless of the method the casino opts for, it's vital that a professional player ensures that the casino's rake does not exceed their own potential profit figures.

We players are essentially sharing the profits with the casino.

If the casino is taking too large a cut then there won't be any profits left for us.

If, on the other hand, the casino is taking a smaller cut, we will make more money.

Step 4 - Learn about the primary goal in Poker and our potential manoeuvres

OUR BODY/EMOTIONS

Our body is something we can use a little at the poker tables, and so it is among our Level One attributes
There is more room to use the body if we're playing live games rather than online. However, the whole body *does* come into play either way, if only very little.
Things like facial expressions are important. Our heartrate. Our body language.
The value of these attributes seems to come from how well we can control them. How well we can control our heart-rate, or our facial expressions. If we are not in control of these things we risk losing loads of money at the poker tables.
I'm not a Zen Buddhist Master Yogi, and I am mostly used to playing online, so I can't teach you how to control your body as well as many might. However, I have been studying Buddhist logic for a while now and I do think I can help -
The body is lorded over by emotions. If we can control of our emotions, we gain control of our body and our mind.

Before studying Buddhist logic I hadn't clarified the difference between *cleverness* and *wisdom*:
Cleverness is the ability to solve problems.
At any given time this ability can be hindered by the emotions that we're feeling.
If we're crazed with anger we won't be thinking clearly so will find it difficult to solve our problems.
If we're overjoyed we won't be thinking clearly either.
The more the emotion affects us the less clever we become. And so:
Wisdom is the ability to not be affected by emotions.
We can separate ourselves from our emotions by ignoring them.

We can use discipline to overcome them, forcing our mind to think through our decisions in an organized manner.
Or we might embrace our emotions and use them to drive us forward.

Through playing Poker I've certainly increased my proficiency with these three techniques
I often separate my emotions from my plays; I might feel bad after losing but that's got nothing to do with my next move.
I often force myself to think through each Level properly regardless of how distressed I may become. (This is very valuable in the early days).
Then there are the times when I embrace the rage and play using the dark side. If a player is annoying me, I study him which enables me to compose a more detailed strategy so as to obliterate him.
Each of these techniques works for me but I'm sure there are millions of other ways to control our emotions.
We're all subject to these feelings throughout our day to day lives and we all have our own ways to overcome them. Perhaps you want to squeeze stress balls, or maybe you can find some way to take it out on your best buddy.
Regardless of how you choose to do it, controlling your emotions will often make the difference between becoming a winner in this game or going home a loser.

My brother has been running a free poker boot camp-style school for many years. I drop in from time to time to offer assistance in his online lessons
Throughout the years I've glimpsed hundreds of beginners' eager to become professionals but unfortunately the majority of these people have now failed.
The reason they fail isn't usually because they aren't clever enough to do the job, it isn't because they lack training or opportunity.
The reason they fail is almost always because they cannot handle

the inevitable emotional strain.

They fall prey to something we call *tilt*.

Tilt happens when your mind becomes incapable of thinking through game-related problems *rationally*.

You lose your concentration due to emotional distress.

If a player is unconfident in their ability, they will tilt often when the luck factor in the game goes against them.

If a player is overconfident in their ability, they will tilt when the opponent out-plays them.

If a player becomes angry, hateful or scared easily, they will tilt often too.

If the player is suffering from cognitive dissonance due to facing higher Level plays or opponents, tilt will follow them like a shadow.

Regardless of the emotional issue which caused the problem, once a person stops thinking rationally while playing any game they will usually lose

These new losses then magnify the emotional turmoil of tilt and the player will find themselves in an even deeper hole which quite often leads to the end of their career.

To contend with this a poker player needs to be well aware that we will all lose in this game very often regardless of our skill level.

To win $100,000 profit, we'd have to lose approximately $2 million!

We wouldn't actually need 2 million to make 100k profit.

We're winning and losing hand after hand all day long and so we'd only need a few thousand in our bankroll if we're playing online.

Tilt does often prevent players achieving this sort of profit, but tilt isn't quite as bad as it seems.

If we're playing a physical game then we expect to develop our physical self by straining through training

It would hurt, but we'd continue on regardless because we'd know that we're getting fitter and more capable of playing.

Poker is much the same but most players don't see it. Instead of grinding on through the pain, many give in.

If we *are* able to work our way around these emotional assaults we learn far more than how to play Poker better.

We learn how to keep our head while under an emotional assault.

We learn about our emotions and how to deal with them.

We learn to concentrate better.

We become wiser.

I've come to realize that every negative feeling which induces tilt has one thing in common: they're all caused by a lack of *understanding*

The angry player can't understand why things have gone against him, he feels hard done by and so turns to rage. The player who despairs doesn't realize that the future will become brighter and so turns to fear. The arrogant player blames everyone else for his losses and so falls out with his coach or starts presuming that everyone's cheating.

This doesn't only make sense in regard to Poker.

In Buddha's eyes, *ignorance is the primary cause of all our emotional distress.*

He says that *if we analyse our problems perfectly they will cease to be problems.*

The best way I know to contend with tilt is to understand all the Levels for our game

It's certainly not a perfect solution to all our emotional problems at the poker tables but it does help massively.

If we know the Levels and then do make a mistake that leads to us losing, we are capable of working out exactly what we did wrong and so won't make the same mistake so easily next time.

In the same way that negative emotions cause tilt, positive emotions help defend against it

If you're a loving person you will feel happy for your opponent when he gets lucky and beats you.

If you're brave you will face your losses with a wry smile.

If you're confident you'll be aware that you're doing well regardless of your result.

Unfortunately, almost all positive emotions do still cloud your mind in a similar way to the negative ones.

They won't make you give up and go back to a menial profession but at any given moment they do seem to interfere with your logical thought process at the tables.

They appear to interfere with our concentration.

The safest way to play is to stay calm. Whether you're currently winning or losing, regardless of the game you're playing, you must remain wise.

When I decided to put emotions in at Level One I did so because they appear to be physical attributes, but we're stepping into that nonphysical 'spiritual' realm again, and so I don't really know how emotions work.

Buddha seemed to get it though.

He said that it's actually the beings wrong understanding of itself that causes it to be distressed.

He seemed to say that our entire understanding of Level One is incorrect at its root.

He described the entire idea of a 'self' as an illusion.

He said that there is no such thing as a self, neither having a self, nor not having a self.

He said that the whole concept of a 'self' is imagined and that this is a prerequisite for all our emotions. (Except for the peaceful compassionate loving feeling you experience upon acquiring perfect concentration).

I don't know if he's right, all I can do is explain it as I see it, and it seems to me that emotions do exist at Level One and are simply unrecognised until we reach the higher levels

To me, emotions look much like everything else at Level One.

LEVEL TWO

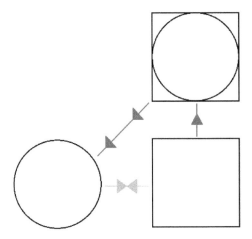

Level Two relates to our perception of our opponent, the "Villain," within the game

At Level Two in Poker we become aware of the opponent. We become aware of *his* physical attributes.
In *Poker* Level Two is all about our opponent, but for this formula to work our opponent could be replaced be any object or any factor regarding anything that exists.
I haven't found anything that we can't consider in place of the opponent.
We can feed allies into this mental calculator.
We can consider our environment or any situation.
We can consider any attribute of any object too.

The opponent, or the *object* being considered, is represented by the square in the diagram
It is our understanding of the opponent that is represented by the distorted square.
The circle, if you hadn't guessed, represents us.

To understand the opponent, we need to consider the strength of his attributes
We need to consider the strength of his hand and the amount of chips that he holds.
At Level Two we then compare his strength to ours to see whether our decision will lead to us *winning*.
We weigh up the strength of his attributes against the strength of ours and make a guess as to whether a specific manoeuvre will lead to us reaching our goal.
Comparing the strength of attributes and predicting winnings is reasonably easy in a game like Poker.
All we need to do is look at our two cards and we'll know which of the opponent's hands we beat. We look to the chips and see how much we might make.
In other games, this is not usually so straightforward.
We can't say for sure whether a football player's stamina is more important than his opponent's speed.
For games like this you need to concoct your own values for the attributes before you're able to consider whether or not you are going to reach your goal.

In Poker, there is only one factor which stops us from being able to calculate our winnings perfectly
There is only one factor in this game which is unknown to us...
The villain's cards
Any hidden attribute in any game brings with it an impossible sum. We can only guess as to its real value.
We can only guess as to the strength of the villain's cards.

At Level Two it is only our understanding of our opponent's

strength that stipulates which manoeuvre we select with our hand

This effect is displayed in the diagram by the red arrows.

These arrows indicate the path of causality from the opponent to ourselves: *The path of cause and effect.*

If we think that the opponent's cards are stronger than ours, at Level Two, we're able to fold.

If we're in a fight and see that the opponent has aimed a punch to our head, at Level Two, we're able to move out of the way.

At Level Two we learn how to defend.

It's only now that we start to learn what to do with ourselves, and so it stands to reason that once a *being* is capable of Level Two *it* can start to learn what to do with *itself*

To be capable of making a Level Two decision the being would first need to be capable of perception, which is not all that easy a skill for a rock to develop.

It would also need to learn how to compare its own strength to that of its opponents, which means that it would have to learn how to utilise *feelings* and *emotions* to separate the value of its own attributes from those belonging to its environment.

The Level Two being might then feel *terror* upon seeing an opponent that is clearly stronger than itself, or it might have to clash horns before it can feel who is stronger.

The ability to do all this would develop over millions of years as the being slowly learns how to best contend with all that nature can throw at it.

The being would steadily evolve, gathering information and developing instincts with only one thing on its mind: *Making decisions based on knowledge gained from past experiences*

Fortunately, we humans already did all of that

When we play Poker we're already aware of the strength of our cards, which is Level One.

We're already aware of the person sat across the table from us, which is Level Two.

To some degree of accuracy, we even know the Level Two calculations which enable us to compare ourselves to the opponent, guess our expected profits, and select a manoeuvre.

And this gives us a pretty awesome ability.

If you don't care too much about accuracy and can't be bothered to learn any of the calculations for Poker, you can still play Poker at Level Two pretty happily provided you keep one thing on your mind:

The villain's range

THE VILLAIN'S RANGE

If we could see the strength of the opponent's cards Poker would be a whole lot simpler.

If his cards were stronger than ours and would continue to be so then we'd usually fold, if they were weaker than ours and would continue to be so then we'd usually continue playing.

But, alas, we cannot see the opponent's cards.

Before we're able to compare our strength to the opponents and select our action we first need to visualise as accurate an interpretation of his cards as we possibly can.

To do this we put the opponent on a range.

At Level One we've already learned how to consider a range. We use the same method to understand how often each hand will appear in the villain's range

As soon as our opponent is given his cards, before he's taken any action, we can accurately put him on a 100% range.

We can accurately say that he could have any two cards.

Then, as soon as he makes a manoeuvre, we're able to start breaking his range down. We can start pinpointing the exact cards that he holds.

By the end of a hand in Holdem, after only very few moves have been made, we *are* sometimes able to say the exact two cards that the opponent holds.

But usually the hand will end before that.

At Level Two the way that we pinpoint his holding is by looking to the past

We use everything we've learned about the opponent, or the type of opponent, to help us better understand what he is doing with his range.

If we see that the opponent is only choosing to play around 25% of hands, and then he plays a hand, we simply assign him a range containing around 25% of hands.

If we've seen that the opponent only ever raises over our bet around 3% of the time we wouldn't expect him to do it with a weak hand. It's only logical to first presume that the more chips that the villain is choosing to put in the pot the better the hand that he's holding. If he raises 3% of the time, we would consider the opponent's range to contain the very best 3% of hands unless we have seen him do something different.

If this same opponent were instead to call our bet then we can use that same logic to remove the top 3% of hands from his range as we expect he would have raised it up with his strongest cards.

Creating a range for the opponent isn't an exact science
Each player will play different cards differently.

We constantly develop our understanding of our opponents and as we learn more about a player our analysis of his range will change.

Beginners often play random hands or make random plays and so for these guys we'd usually give them a hazier range than we would a Reg.

We can't say precisely how a beginner will play a certain hand, but then, he probably wouldn't be able to tell us himself either.

Beginners don't fully understand what they're doing at the tables and so they are inherently unpredictable.

This is bad for us, but these guys make up for it by rarely mixing up their plays. They make the same plays with the same hands over and over and over again.

If we see a beginner raise the minimum amount one time with the very best of hands, the chances are that the next time he *min-raises* he'll probably have the same sort of hand.

Against these inexperienced players predicting their range leads us to the greatest profits we'll ever find at any poker table.

It's very valuable to watch closely for any of these consistent

betting patterns from any players
We have to work out what they are doing with their cards!
The more accurately we understand the strength of their hand, the more accurately we can select a beneficial manoeuvre.
And don't forget, if we know what the opponent does when they are strong then we can tell when they are *not* strong.

If possible, we want to be making notes anytime we find out any information on a player

Each individual player could have plays they make often, plays they make rarely and plays that they never make:
If we see that the player doesn't make a certain play at all, perhaps he never raises over our bets on the flop, then the player will most probably have a blank spot in their knowledge of the game.
Blank spots like this make it far easier for us to play against them as they have less moves in their artillery.
This is the same in all games as it is in Poker.
If a boxer never uses his left hook then we could constantly side-step to the right without any fear of being hit.
If a poker player never raises our bets then we can always bet against him without bothering to consider what might happen if we get raised.
This doesn't mean that we always benefit if the opponent leaves plays from his artillery. He might be choosing to remove certain plays deliberately which does make it more difficult for us to pinpoint his strength.
If he never raised over our bet then we will never know whether he has the best kind of hand or not and so we won't be able to remove that 3% of his range which I just mentioned.
It is usually easier for *him* to play if he's making fewer types of plays.
It can also massively confuse and frustrate us if we keep seeing him make the same moves with different cards. And when we're confused and frustrated, we make less profit.

If the opponent is instead the kind of player who makes a specific play more often than could make sense, we do what we always do
We change our understanding of their range accordingly.
Perhaps the opponent has been betting on every flop?
Perhaps he checks every flop?
Perhaps he calls every time we bet on the flop?
If the opponent were doing any of these three things then the range we gave him pre-flop would be identical to the range we gave him on the flop.
If we expect him to make a specific play with his entire range then it stands to reason that his range would not change when he makes that play.

We can also use his other attributes, like his chips and his position, to help us understand his hidden qualities.
If he's aware that he's in a weak position, we can assume that he may well know to play with stronger cards than usual.
If he's in a stronger position, he will probably play more hands than usual.
If he's about to run out of chips when the blinds reach him, we can assume that he may well play with any half decent hand.

The most beneficial thing to learn about an individual opponent is his overall understanding of this formula
If the player has no awareness of a certain aspect from any Level then we will find it far easier to put him on a specific range of cards.
We will also know which plays the opponent is not comfortable defending against.

His attributes, his betting patterns, his characteristics and his understanding of the game all help us make reads on his range
What we're doing here is building knowledge about a certain player or type of players to help us select our manoeuvre.
But if we really want to know what the opponent is doing, if we

really want to understand the opponent's range, we need to know *why* he does one thing or *why* he does another.

We need to understand *why* he makes his plays.

We need to find the reason behind Level Two, which is found at Level Three.

Using only Level Two we can put the villain on a precise range by watching the game like a hawk

We aim to understand anything and everything about him.

We look to his betting patterns, his personality and his understanding of the game to find out as accurately as possible what it is that he does with the different parts of his range.

Does he always play his strong hands slowly?

Does he always bet big with his biggest hands?

Is he fully aware of Level Three?

Every piece of information we gather on them helps us decipher their range.

Step 5 - Consider the strength of the opponent's hand by putting him on a range.

HIS BOARD

As soon as we're able to concoct a range for the opponent we're able to see how often that player has hit specific boards

Let's presume that we're up against a low stakes Reg who's called our opening bet pre-flop. We see that he plays 7% of cards when he makes this call, and that he would have raised with 3% -

The range I'd give this person, without any further information, would contain mostly pocket pairs from 22 - JJ which makes up around 5% of all possible hands. I'd also expect him to hold big broadway cards too, AQ, AJs. So, let's say that these two hands make up the other 2% of his range.

Now try this yourself. Try to consider how often that range hits on these three completely different boards:

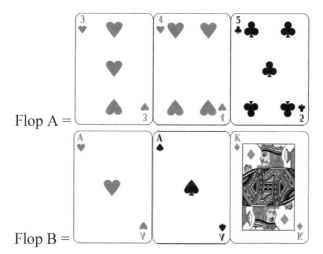

Flop A =

Flop B =

Flop C =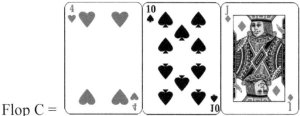

- The first of these flops smashes the villains range. *All* his pocket pairs have either hit the board to become sets, are over-pairs (pairs bigger than the board), or are draws with a good chance of improving, and a quarter of his suited cards have a good chance of hitting a flush. Even his unsuited cards are over-cards and so he also has a chance to pick up a big pair on a later street too.
- On the second flop, we can see that the villain most probably hasn't hit. There are two Aces on the board, so we can remove those cards from his potential range and can see that he has far more pocket pairs that have missed rather than Ax hands that have hit.
- On the third and final board, we can see that the majority of the opponent's range has completely missed. All we really need to worry about is the rare 44, TT, JJ and AJs which could have hit.

As we don't know *exactly* what it is that the opponent holds, there isn't much point in working out *exactly* how often his range hits a specific board. When I see a board I roughly consider which parts of the opponent's range has hit or missed and this is usually enough information for me to select my move.

Once the board has arrived it's easy enough to see how often the opponent has hit, but to really understand the strength of the villains range we need to understand how often the different future cards will strengthen their hand

To do this we use the same sums we learnt at Level One, except now we calculate the chance for all the different hands in the villain's range improving.

THE LEVELS

Board =

This time let's say the villain is seated in an early position and that he has bet against us both preflop and on the flop
We're likely to have assigned a strong range to him by this point. So, let's say that -

- A little less than half of the villain's remaining range is KQ.
- A little less than half is AJ.
- He also has the odd TJ and Set in there too.

Now we need to consider the chance that he will improve to a straight, to a single pair, two pair, to a full house or quads perhaps.

- If the villain has KQ then he has 8 outs which lead to him making a straight. There are 6 cards known to us, so there are only 46 cards left in the deck. This gives him around a 1 in 6 chance of hitting the straight. If he does have KQ then he would also have a chance of hitting a K or a Q on the river which would give him a pair, to hit a pair he has 6 outs and so around a 1 in 8 chance. We then add these together and see that in total, the KQ has almost a 1 in 3 chance of improving.

- If he has AJ then he has 6 cards which will improve his hand and so has a 1 in 8 chance of improving.

- If the villain has TJ already, then he has 4 outs to the full house and around a 1 in 12 chance of improving.

- If he has a pocket pair which has turned into a set, then he has 10 outs and a 1 in 4.5 chance of improving.

These sums may sound difficult, but when your sat at the tables you rarely need to calculate them accurately
You may notice that these are the same sums from Level One reflected here, so there isn't much more for you to learn.
Each new Level works in the same way. Each new Level is just a reflection of the one before it which adds a little bit more to our calculation.

Step 6 - Consider how strong the opponent's range is on certain boards, and how strong it will become in the future.

ADDING THE CHIPS

Once we understand the strength of the villains range we can compare the strength of his range to the strength of our hand to see how often we expect to win.
This is called our 'win-chance'.
Before we can select our play we still have one more thing to consider - We need to add chips to the equation.
We need to find out how many chips we will make on average against the hands in the villain's range. We need to find out whether our play will be profitable. Only then can we see our decision will lead us to achieving our goal in this game. Winning chips.
There are many ways to make calculate the coming sum, I will start with the most difficult and most accurate equation and then I'll show you easier ways of doing it.

To keep things as simple as possible, let's expand on the example in the last chapter so that I can show you exactly how we calculate what is known to the Poker world as 'value' -

Board -

The villains range:

- KQ – slightly less than half of his range
- AJ – slightly less than half of his range
- TJ , Sets – around 1/10 of his range

Let's say that our cards are:

And that we're facing a bet of 100 chips into a pot of 100.
Let's also say that the opponent has bet all his remaining chips so that we're deciding between just two different plays, calling or folding.
We need to find out if there is any profit in a call, and if there is not, we need to fold
To find our expected profits we need to periodically go through each hand in the villain's range and work out how much we will win or lose against each on average, we then add the numbers together to find our average profit.

Let's start with KQ
The total pot is going to be 300 and we will win 2/3 of the time, so, against KQ we will win 200 on average. As it will cost 100 to call, our profit will be 100.
And so... KQ = +100

Against AJ
The pot will be 300 and we win only win around 1 in 15 times so our winnings is about 20. As it will cost 100, our total losses are around 80.
AJ = -80

Against sets and TJ we lose every time, so our loses are 100
TJ/Sets = -100

Before we can add these numbers together we have to account for the fact that some hands will appear more often than others
We guessed that AJ and KQ both appear 4.5x more often than the other hands, so if we can give them 4.5 instances for every 1 instance of TJ/Sets we're able to calculate our average profits by dividing the total number by the total number of instances.
The sum looks like this -

KQ... 4.5 x 100 = 450
AJ... 4.5 x 80 = - 360
TJ/Sets = -100
450 - 360 - 100 = - 10
-10 divided by the 10 instances gives us an overall profit of -1.

This play is pretty much breakeven
Remember, we don't know how much of the villain's range is really KQ, we just guessed. Our profit number is not going to be perfectly accurate. It is just an estimation.
Using this estimation we can see that for this decision it wouldn't really matter whether or not we call.

When we're sat at the poker tables there isn't much point in calculating an exact profit number because we don't know exactly what the opponent holds
The calculation I have showed you is the easiest way to calculate an exact number for the profit from this play, but as we're never going to be accurate anyway, we don't really need to put a *number* on our profit. We just need to know roughly whether or not we are likely to make profit from the play.
There is another calculation we can often use which is a far easier way of calculating the plays profitability. We can simply compare

our win-chance with the odds given to us by the price (our pot odds).

In the example, we can see that we are currently beating nearly half the opponent's range. We are currently beating all the KQ hands, but, 1/3 of those KQ hands will improve to beat us which leaves us a total of around a 1 in 3 chance of winning.

And thus, we have our win-chance.

The size of the call in relation to the total chips we stand to win gives us a price around the same. 2:1.

And thus, we have our pot odds.

(The difference between probability and odds is ridiculous and so I feel I should clear it up for you. The probability being 1 in 3 is exactly the same as the odds being 2:1. Probability is: chance for and *total chance*. While odds is: chance for and *chance against*... Who makes this stuff up!?)

All we have to do is compare these two figures and we can see that there is not really any profit in the call, but nor will we lose anything either. This call is somewhere near a break even play.

Both of these methods are important to be aware of. They will both serve you at different times. But you don't need to accurately calculate these sums at the tables

It'd be a minor range related factor which swayed my decision on which manoeuvre to make on a close decision like this

If I thought there's a small chance that the opponent has some kind of bluff in his range then I'd call. If I thought the opponent might have some small chance of holding JK or JQ then I'd fold. Regardless of my play I'd not let myself get demoralised if I did find that I'd made a mistake in this instance.

This was a close call and so even if I do end up losing it really doesn't matter very much. On average I won't be losing much money by making the wrong play in any situation where the numbers are so close.

Step 7 – Select your play by either calculating the chips you will win/lose from the hands in the villain's range, or, by comparing the price of the play with the win-chance *(If the price is low in relation to the total winnings available, you need not beat much of the villain's range in order for your play to be profitable.)*

IMPLIED PROFITS

There is one important factor which I left from that last example: I made it so that there were no further betting opportunities available so that we could disregard the *Future-Present*.

The only time we win or lose chips is when the hand reaches conclusion
At Level Two we don't *actually* need to consider our profits (value) for any given instant. Instead, we need to consider how much value we stand to win, or lose, by the time the hand reaches showdown (when the cards are turned over to see who wins).

To find our profits come showdown we need to consider our *implied profits*
Let's go back to that last example, except this time let's say that the opponent didn't put in all of his chips. Instead, let's say that both of us had an extra 1,000,000 chips behind.
Now that there is loads of extra money active in the pot we can't easily guess how much we will win or lose in the end. There is no way we can tell how big the pot is going to be by the time the hand reaches its conclusion. All our potential profit figures become confused because they depend on actions which haven't happened yet.
When we start considering implied profits we find that this unknown factor presents a real problem. We won't know how much money the opponent is going to put in the pot until after we see them do it.

So, we do what we always do... We just guess
Is he the type of guy that will never bet on any further cards regardless of his holding? Or maybe she's the type of girl who'll go mental and chuck in all her chips every time she misses?

To calculate our implied profits, we guess what the opponent will do in the future with the different hands in their range and then we adjust our potential price or potential winnings accordingly.
If you're a beginner, I wouldn't worry about this stuff yet. Considering future plays and implied profits isn't too important. Provided you're able to understand whether your current play is profitable against the opponent's range you won't find that the future decisions will bring along very many changes that will turn your potential winnings into potential losses.

Step 7.1 - To find your final profit figure you need to consider how many chips you will win by showdown.

THE EASIEST WAY

We can further simplify our play selection, but by adopting this technique we do lose quite a lot of accuracy
What we can do is forget about the chips entirely and concentrate only on the cards.
The most common phase at which we might adopt such a rudimentary technique is pre-flop. While we're at the pre-flop stage of a hand the opponent's range is so wide, and the potential futures so broad, that to find the value of our play by using either of the methods I described would be impractical.
Instead, we could just select to play hands that are stronger than the villain's range.
If we expect that the opponent has AT+ then we don't play any cards weaker than AT against him.
We might choose to only play AJ+, or if we want to be more conservative, we could go higher and play only AQ+.
If we do select our plays in this way whilst pre-flop, on average the board will make us stronger than the opponent.

At the pre-flop stage of the hand it is extremely valuable to consider the chips we might win or lose against the different hands in the villain's range depending on the different situations that may come and the lines that we both might take
Considering the chips and the future moves etc is still massive.
The only time the strength of our hand really matters is when the cards are revealed at the end of a hand and so it could be deemed all important to keep the profits made at showdown in our mind.
I personally do try to make as detailed an estimation of the entire play as I can before I select any pre-flop action.
This is a rather advanced skill, which requires the grouping of different futures into possible winning and losing futures and so I

wouldn't advise you to worry about it until you are comfortable with all the other factors you need to consider.

I usually advise players to simply attempt to stay ahead of the opponent's range while considering what to do before the flop comes down at Level Two.

If you do decide to adopt this rudimentary technique for selecting pre-flop cards, then you will need to work from standardized bet sizes

If I could choose one standard bet size for all pre-flop actions it would have to be pot size.

To cater for all the equations that we're forsaking we need to then change our standard sizing depending on the kind of situation that we're in:

If we're in a good seat, we'd lower the bet size slightly.

If the opponent has a small stack containing less than 100 big blinds, we'd want to lower our sizing slightly to account for the implied profits modification that we didn't bother making.

If the opponent has a big stack containing more than 100 big blinds, we'd want to raise our sizing slightly for the same reason.

There are infinite reasons which will lead to change your bet sizing

As you develop your understanding of the game you'll begin to find more and more reasons to adjust your standardized sizing.

However, you will also find it easier to consider future profit figures against the different hands in the villain's range and so these standardized sizes become less important as you get better at the game.

In case you're wondering, if I could choose any standard bet size for post-flop situations, I would choose 2/3 pot.

Step 7.2 - Whilst the future is at its most vast, comparing your price to your win-

chance may not be a practical way to select your play. And so instead, you could forget about both the chips and the future cards and concentrate only on making sure that your hand is stronger than the opponent's range.

THE VILLAIN'S BODY/EMOTIONS

Lots of professionals deliberately attempt to put their opponent on tilt
They try to make the opponent lose their mind so they themselves can make extra profit.
I am very much against this practice. Especially when I'm playing against people who are clearly less skilled than myself. Not just because it's a nasty thing to do, which it really is. The main reason I'm reluctant to put my opponent on tilt is because I want the weaker player to come back and play against me again.
I'd much prefer that a player loses to me with a smile on his face than to have him walk away hating the game and swearing never to play again.
Yes, I could probably make more money on a specific day if I make him go crazy, but on another day, I would make far more money if the weaker players came back.
Having said that, if I ever find myself facing a guy who gives everyone an ear-full himself, I do feel justified in destroying his moral.
I myself don't want to play against somebody who tries to drive me insane all day, and I don't want them ruining the experience for the weaker players either, so I'll happily make these guys leave and never return.

It's not often that I attempt to induce a negative feeling from the opponent, but I do often attempt to understand how it is that my opponent is feeling so that I can pick up physical reads to aid me whilst breaking down their range
Lots of players say that you can get reads on your opponents by looking at them to see how they're throwing their chips into the pot, or by watching their eyes to see if they become dilated.

It *is* ridiculously profitable to find a players *tell* so do keep an eye out for them, but the finding of these tells usually comes as a surprise and so they aren't worth spending your time searching for.

I much prefer to look at my opponent to try to see how they're feeling. This is something we can always be aware of at the tables. Recognising a person's emotions is a skill that we all use every day.

We already have the ability to tell if the person sitting opposite us is happy or sad, and that can be of huge benefit while facing them at the poker tables.

If the board comes and we see the opponent looks relieved, then we can naturally assume that he's hit.

If the board comes and the opponent looks angry, we can guess that it's likely he's missed.

If the opponent becomes excited when that flop comes down, we'd guess that he has something strong.

Some guys will become so excited that their hands shake, but be careful, some guys become so scared that their hands shake, and some people just have shaky hands.

If we're playing on the internet we can't see our opponent and so taking physical reads becomes more difficult, but it's not completely impossible

In a similar way, we can use timing-tells to help us read the opponents holding.

Some players will rush their play when they are bluffing, some will take their time when they are strong. They might make a very quick 'snap' decision whenever they are very weak, many will do this same thing when they are very strong. Lots of players do this when they are in a pre-planned situation.

As usual, it's our job to keep a little eye on any timing-tells so that they can help us put the villain on a range.

All information we can assume to any degree of certainty aids us when predicting our opponents holding!

Naturally we're able to pick up the odd read on our opponents by becoming familiar with their emotions or trends, but casual players sometimes tell us all about their style of play without us prompting them in any way.

In small games at live casinos the players often spend their time talking about their own understanding of the game. "I'd always fold with that hand" or "I'd go all in if I was him just then".

This is a really big mistake.

I *would* certainly advise you to talk to your opponents, even if you're only aiming to become familiar with their personality and emotions, there is almost always a lot that you could learn from them

You might eventually find that you have to choose between your friendship and your win rate, but that's a decision for you to make on your own.

LEVEL THREE

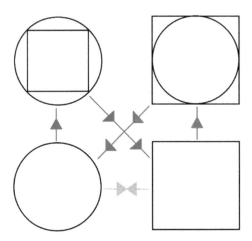

Level Three relates to our opponents understanding of us, within the game

Level Three requires empathy.
Normally when people speak of empathy they speak only of feeling what another person is feeling. At Level Three in the decision-making formula it's all about the understanding side of things. It's all about understanding things from the opponent's perspective.
To do this you need to put yourself in the other person's shoes.
You need to become the opponent. See what they see. Feel what they feel.
Once you can do that, you can predict how the opponent will react to outside influences. You can predict how they will react to you. At this Level in Poker we see how it is that our own actions will

affect the opponent. We see how our actions will affect the villains range.

If you think back to Level Two, you should remember that our understanding of the opponent stipulates how it is that we manoeuvre (the red arrows).

Well, it's the same for him as it is for us. He makes his plays based on his understanding of us (the blue arrows).

That new shape on the diagram, the distorted circle, represents our perceived attributes. The cards that the opponent thinks that we hold. Our perceived range.

What we do at Level Three is manipulate the opponent's perception *of us* so as to lure him into making a play which is more profitable *for us*.

We deceive our opponent for extra money.

By controlling the way that we are perceived, we gain control over what it is that the opponent does.

Controlling our opponent is extremely valuable but it gets much, much better

Most poker players are stuck at Level Two and are incapable of comprehending theory related to Level Three.

I gathered over 40,000 views over a couple of threads that I had placed on an old poker forum. In those threads I described aspects of Level Three very clearly.

A clever man from the community at that website even helped me explain it.

Yet even as I finalize this training guide over three years later, the vast majority of people at that site still refuse to accept that Level Three makes any sense.

This clever man and I explained everything in amazing detail.

We gave clear examples.

We showed exactly how to make Level Three plays and explained why.

We were even teaching for free.

But, alas, these otherwise smart individuals couldn't understand what we were saying.

Instead of giving us thanks they became abusive.
They got really, really angry.

These unfortunate fellows had developed powerful mental blocks. Cognitive dissonance was obliterating their minds
They had almost entirely lost the ability to recognise unknown higher-level poker logic.
If we can kindly point out all the factors relating to Level Three, if we can wave the information in front of the faces of these players without them recognising any sense in what we're saying, imagine what chance they stand of working out what we're doing when they can't even see our two cards!
At the tables we destroy them over and over and over again.
Low Level Regs have screamed at me for years, "You're a fool!! You should never have had those cards!"
Sorry my friend but you're mistaken. I deliberately made you *think* I didn't have those cards. I planned the future of my perceived range and I expected that I could make more money from your range by making a disguised play rather than a straightforward one.
You really can explain to them how you made the Level Three play if you like. They won't understand.
They won't comprehend it. They can't.
They'll just go mental.

To be capable of empathy a being requires the development of a new ability – *the ability to create in their mind an image of a situation which they have not seen*
A Level Three being is able to consider 'fake' Level two situations. They're able to *imagine*.
Using this one ability beings are able to imagine what their own attributes look like from the outside, and thus they acquire self-awareness.
They learn to imagine things from the perspective of the object or opponent, and so acquire empathy.
They're also able to imagine their own future self.

These three abilities look almost identical to me
I don't have any solid reason to say why they all come in at Level Three but I would bet a whole lot of money that they do.
This is one of the very last pieces of the puzzle for me and it all fits in perfectly.
The only one of these abilities that I can pin to Level Three is empathy.
Self-awareness is clearly Level One and seems to replace the need for the feelings and emotions that were essential to a Level Two decision.
Planning the future, I presume, comes from Level Three. It does crossover into Level Two. At Level Two we do still need to consider the future, and the opponent will be considering the future when choosing what to do with his range.
But *all* Level Three factors cross over to Level Two in *some* way. They all cross over to Level One too.
The entire purpose of Level Three is to enhance our understanding of Levels One and Two.
Empathy, self-awareness and future planning seem to be the tools we use to achieve this heightened level of understanding.
Empathy, and the control that comes with it, really is awesome, however, I can't help but think that the primary reason we developed an imagination must have been so that we'd be able to imagine different potential futures.
At this stage of a being's development it'd no longer be forced to make decisions based on past experiences.
At Level Three we become capable of making plans.
We can now plan our offensive.

We have two separate types of offensive plays available to us
We can either confront the opponent in a straightforward manner (Level Two attack), or we can deceive him with a disguised attack (Level Three attack).
Obviously, we *can't* disguise any attribute of ours which is visible, like our chips at a poker table.
We can only disguise attributes which are hidden.

In Poker, we can disguise the strength of our hand.

The ability to make a disguised play is obviously hugely important for a poker player, but it's equally as important in most other games too
In Football, a player might fake left then go right. He might step over the ball multiple times to confuse the opponents understanding of his plan.
A bowler in a cricket match might curve the ball or he might throw it straight.
All the same logic applies.
The Levels work the same way in Poker as they do in *all* games.
If the opponent has not memorised a strategy to counter our disguised play and he's not comfortable with the Levels then he will experience cognitive dissonance in some way when confronted with the disguised play and so will often make a mistake.

It's important to be aware that it is only useful for you to deceive your opponent if it coincides with the final goal in the game
There is no point in curling the ball if the opponent can't hit the straight throws.
It would only be worth your army appearing where your opponent didn't expect if it was going to help you win the war.
Making a business move your competition doesn't expect is only a good thing if it's going to create more profit for your company.
In Poker, the final goal is to make more chips and so if it would not make us more chips from the opponent's range by disguising our cards then we would not do so.

All we need to do is look to the potential profit and see which play we think would make us more money off the villain in the end
We do the same as we've always done.
Level One and Level Two have not changed.

All we've done is recognise the reason behind Level Two.

We recognise that the opponent's actions are based on his understanding of our cards. That the opponent is capable of Level Two.

We've become aware that our actions affect our opponent and now we can pre-plan their reaction to our manoeuvres.

We've gained a broader understanding of the game, and so more options have become available to us.

Sun Tzu, a famous military general of old, clearly recognised the importance of his image on the battlefield

He often said things like, "Appear where your opponent will not expect".

He also seemed to back up my logic with this quote:

"There are only two forms of attack, a direct attack, or an indirect attack."

He was clearly extremely talented when it comes to Level Three in respect of war. However, I can't help but think that his most famous statement was wrong.

Sun Tzu once said:

"All warfare is based upon deception."

I would love to ask him how a direct attack is based on deception. A direct attack wouldn't be deceptive at all. It's actually the opposite of deceptive. Right??

Level Three manoeuvres *are* all based on deception. The direct plays that we're able to make once acquiring an imagination are Level Two plays. When making a direct offensive we disregard the information related to our image (our perceived attributes) and act solely on Level One and Level Two values.

When we make a straightforward play, we do first have to consider a deceptive play, so perhaps that's what Sun Tzu meant. It's hard to be sure.

There is a chance that I completely misunderstand the great strategist in regard to this statement, but then, I have come to suspect that he puts too much stock in Level Three...

OUR PERCEIVED RANGE

Playing at Level Three is very easy for us human beings.
Our amazing subconscious happily takes care of all the hard work for us.
In Poker, all we need to do is follow our *perceived range* and the subconscious will happily do all the rest.
To be accurate though, you do need to consciously understand exactly how this new Level works.

Every single time our opponent makes a decision he does so due to his understanding of *us*
If he thinks our cards are stronger than his (in relation to his price) he folds. If he thinks our cards are weaker than his he continues playing.
If we can predict how the opponent's perception of our range will change based on our manoeuvres, we can see how to lead him into making the decisions which *we* want him to make.

Our perceived range is much the same as any other range

- Our past actions are still key to the creation of our perceived range

- In the future, we still find the results of our actions on our perceived range

- And the chance of our perceived range improving is still the same

We only really add one new piece to the puzzle - We are not considering the cards which we think we would have. That would be our Level One self. Our actual range.

At Level Three we need to consider ourselves from the *outside*. We need to consider the cards that the villain thinks that we have. We need to see the game through the villain's eyes.

To do this we first need some understanding of the opponent's ability

We need to know whose head we're climbing into before we can understand how we look from their perspective.

Once we've established, to some degree of accuracy, the opponents understanding of the game we can start considering how he may interpret our actions.

We know that he uses the past to make his judgment on us and so that is how we find out how he perceives us.

If the opponent has seen us raise with a weak hand he will probably start to think we raise while weak.

Less experienced players will take much more time to recognise our plays. If an absolute beginner only ever witnesses us raising with the very strongest of hands then we need not worry about it because he probably won't have noticed.

Beginners usually consider far fewer things than a regular player. They *will* probably notice if you suddenly make a big bet after being asleep for 30 minutes, but they probably *won't* notice if you only ever raise over bets when you hold a strong value hand.

Regardless of how much information the opponent considers about us, we can always translate it into a perceived range for our self

We must always follow this perceived range in our minds in order to maximise our profit. There is no time when it wouldn't be important.

Even if we don't eventually choose to mislead the opponent, even if we choose to make a straight forward play, we wouldn't have known whether it would've been more beneficial to make a disguised play unless we'd first imagined our own perceived range and how it might change based on our action.

We break this new range down in much the same way as any other range
If we've only just received our two cards and have performed no action, we can accurately say that our perceived range contains any two cards.
Every time we make an action from this point onwards we break down our perceived range making it more and more precise.

At any given time, a beginner could centre on one single hand for us, they 'Put us on AA'
We can certainly take advantage of this kind of conscious assumption, but we must also be aware that in truth they are subconsciously considering an entire range for us.
Deep down inside they know that there's a chance we could have all sorts of hands. What their subconscious is really trying to say is that AA is a dangerous part of our range.

A skilled player does usually take more into consideration than a weaker player, but if they are unaware of the higher Levels, which most are, this skill actually gives *us* a massive advantage over *them*
The more accurate a range a Level Two player puts us on, the more accurate a perceived range we can ascertain for ourselves.
We then become capable of pretending to show him that we have exactly the nuts, or exactly a bluff.
We have much, much more control over an accurate Level Two player than we do a beginner who trusts in his instincts.
The accurate Level Two player only really has one way to battle back against our abuse - He'll learn how we act in individual situations and then in those specific situations he'll take his own play to the next level.
He'll memorise higher level strategies and plays.
When the Level Two opponent witnesses what we do in an individual situation they'll adapt their understanding of our range to take advantage of their newfound knowledge.

THE LEVELS

It's our job to recognise in which situations each individual has realised that we're abusing him in.

Once we see that he has made a change to his range to react to our plays we adjust our understanding of our perceived range accordingly.

If he thinks we often bluff him in a specific situation, we simply add more bluffs to our perceived range for that situation.

Easy as pie.

The potential for a counter attack from the opponent might seem like damning evidence against the profitability of Level Three

If you think that this is the case, I can almost guarantee that are an experienced player who has not yet advanced beyond Level Two.

You're yet to unlock Level Three in your mind.

There are trillions of situations in Poker and we tangle with individual players reasonability rarely. They can see what we're doing even less often.

Learning to defeat a Level Three player one situation at a time is like climbing a mountain taking one baby step every Christmas.

We're even able to counter our opponent's defence by creating a heavily false image for ourselves

We can spend some time making plays in an extreme way, perhaps by raise bluffing all our weak hands.

We then watch closely and wait until the villain has recognised what we're doing. Then the moment he adjusts his play to contend with us, we make a switch over to strong hands.

We pre-plan not only the way he will adjust his play to contend with us, but also the way we re-adjust against him.

We can apply this kind of strategy over and over again.

There's nothing stopping us from deciding at any given time to play aggressively with our weaker hands.

We can play around with our image forever, constantly manipulating the opponent's understanding of our range so that we can constantly take advantage of him.

His Christmas baby steps become the move we use to annihilate him *even* more efficiently.

Step 8 - Consider your perceived range.

BLUFFING

Bluffing adds an entirely new revenue stream to this game for us. Poker is not only about winning chips from our opponent by having a stronger hand than his.
This game is not only about winning *value*.
Every time the opponent folds his hand we make money as well. At Level Three we're able to pretend to hold a stronger hand than our opponent to get folds from cards which are stronger than ours. We can bluff.

The additional maths equation is very straight forward provided you understand how to calculate profit from value
If we were to bet the size of the pot and were bluffing with no chance of actually being stronger than the villain, our perceived range would need to appear strong enough to fold out at least half of the villain's range for it to be a profitable bet.

In order to bluff effectively we need to pretend to use cards that are in our perceived range
If we pretend to hold cards that the opponent has already disregarded as a potential holding for us, he will not believe the bluff and will not fold.
If our perceived range contains *lots* of hands which have beaten the hands in the villain's range, and then we bet, our bet will look strong and the bluff will likely be successful.

By now you might see that Poker is a battle of ranges
At Level Two it is our cards vs the villains' range. But by Level Three we realise that our cards aren't all important. The real battle is our perceived range vs the villain's range.
If we can control our perceived range we can control the

opponent's range by making him think he needs to fold when he doesn't.

Our actual cards are only important once the hand reaches conclusion, and so until then, it's all about making chips from our perceived range.

When we come to consider how the future effects our new revenue stream we find that we do sometimes benefit by making the opponent fold hands which are currently *weaker* than our holding

We'd do this to stop his hand improving to become stronger than ours in the future. Essentially, we're bluffing against his potential strength come showdown.

This type of play is called 'protection'.

Protection is one of the least significant factors in Poker, but unless we hold the very strongest of cards there is almost always some part of the villain's range which could improve to beat us, and so protecting is almost always relevant during any poker decision. However, making a play *purely for protection* is quite often the wrong move to make. There is almost always something better that we could do.

The common definition of a bluff is:

"An attempt to deceive somebody into believing we are doing something when we are not".

Therefore, we don't only bluff by pretending to be stronger than we really are. We can also pretend to be weaker than we truly are to get more value from hands which the villain might not have usually continued with.

Notice that at Level Three we are always attempting to represent a range which is as far from our actual hand as possible. We try to make ourselves look either stronger or weaker than we truly are.

And that is pretty much all we do at Level Three.

Before making an action we consider how it will make our perceived range look so that we can drag our perceived range as far away from our actual hand as is *profitable*.

If the opponent knew the cards that we are using, or the actual range that we are using in any given situation, he would play perfectly against us.

And so we deceive him, to make money.

It's usually the lower end of our range which would be worthwhile disguising and it's usually when we are at our strongest that we would prefer to play in a straightforward manner

If we have KK and once we make our play our perceived range will contain 55 – AA, there's little point in attempting to play slowly to lessen the KK's standing in our perceived range because the villain will already play many weaker hands against us.

If you think about it, our hand is pretty well disguised anyway. (Perhaps that's what Sun Tzu meant).

We have many weak hands in our perceived range and so the KK is still very strong in comparison to all the cards the opponent thinks that we could have.

If instead we held KK, but then once we make our play our perceived range would have only AA and KK in it, it would usually be more valuable to make a disguised play instead. If our hand is at the bottom of our perceived range the opponent is only likely to continue with something better.

Step 9 - Consider how the opponent will read the changes to your perceived range after you take your action. Will you look strong or will you will you look weak? Will this make you more money than a straightforward play?

PRE-ADJUSTING

In all games, there are two ways we can deceive our opponent:

1. We can bluff by considering imminent changes to our perceived range. We can pretend to hold cards that are, at present, in our perceived range which we don't actually hold... In war, we feign to go one way and then suddenly change the direction of our attack.

2. The second way we can deceive our opponent is by utilising the future present of our perceived range. This is called *Pre-Adjusting*.

We induce the opponent to change his understanding of our perceived range so that during future decisions we can hold a hand that is *not* in our perceived range.
In war, we slyly position our troops in a place where the opponent won't expect, planning to reveal them at the opportune moment.
In Poker we make a play which induces the opponent to remove the hand that we truly hold from our perceived range so that after future changes have materialized we can take advantage of his wrong assumption.

In short, we mash them up
If I have a flush and the opponent thinks to himself, "Yadi could have a flush", then he will play cautiously against me.
If I have a flush and the opponent thinks, "Ahh, I already learned that Yadi does not have a flush!" Then I will decimate him.
He might try to represent a flush himself and raise big in an attempt to make me fold whichever mediocre hand he thinks I hold. He might call me light, thinking that I'm bluffing when I raise him for my entire stack.

Making the opponent think that we could not possibly have hit is hugely valuable when we hold a strong hand

In the same way, if we've completely missed the board but the opponent thinks it smashes our range, it becomes very easy to convince him to fold.

If we've hit a medium strength hand, and the opponent thinks we have only a very strong hand or nothing, he will often leave us to control the size of the pot.

One way or another, if the opponent distinctly believes that you've already folded the hand which you truly hold, you can make, or save, lots of extra money against him.

Pre-Adjusting is a far more effective way of avoiding our perceived range than bluffing, but it is also much harder to do.

To successfully pre-adjust we need to consider what our perceived range will be during our next move and the profit we might then make. This means that we need to consider the future flops, turns or rivers and consider how our perceived range will look on these cards as well as the actions that both we and the opponent will make and how much money we will then stand to win.

We need to consider the future-present.

We need to consider how everything will look during our next move.

To make a pre-adjustment post-flop is both easier to do and less effective than pre-adjusting pre-flop

After the flop comes down it's sometimes worthwhile to make an action which will induce the opponent to remove the strongest parts of our actual range from our perceived range, but this is rarely the best thing to do pre-flop.

Our range is very wide at the early stages of a hand. Our strongest cards will still be relatively well disguised, so it will usually be more profitable to play our strength in a straightforward manner.

The most valuable cards to remove from our actual range before the flop comes down are the weakest of those the opponent expects us to hold:

We can accurately say that the weaker part of our perceived range wouldn't make much money by being stronger than the opponent. So instead of actually playing those cards we replace them with hands that the opponent won't expect.

This gains us far more value, and by removing the weakest of the hands that the opponent expects us to hold we even maximise our profit from bluffing too.

The best way to understand this is to work through a problem using an example:

A common Level Two re-raising range preflop might contain:

A2s – AKs
A9o – AKo.
KQ
High Pocket Pairs.

In this day and age this is the kind of range that a Level Two villain might expect us to hold so it becomes our perceived range against a Level Two player.

Now consider this second range
It's approximately the same size as the first but it'll hit less often... Think about which one you would prefer to hold when you are re-raising this Level Two player:

A5s - AKs
ATo - AKo
KQ
56s - 9Ts
High Pocket Pairs

A Level Two player will usually believe that the first of these two ranges is a better holding. (Unless he's memorised a different strategy for this situation). We're stronger and so we're in a better place. Right?... I'm sorry, but this is wrong.

Remember back to your Level Two lesson about hand selection. The opponents will only be aiming to continue in the pot with hands which are better than ours which makes our weak Aces almost worthless. The opponents will not be continuing with A2 against our perceived range. He will have already folded his A2 pre-flop because he wants to stay ahead of our range!

The main way that we'd make any money by holding a hand like A2 is by getting the opponent to fold. And if that is the case, we needn't actually hold those cards!

If he's to fold then we'll not see a showdown and so we don't actually need to hold that hand.

If the opponents will never see that we held A2 there is no point in ever having it.

We might as well fold our A2, and replace it with a hand the opponent won't expect. We might as well pre-adjust.

If you think back to earlier lessons, you'll also realise that because we don't actually have A2 in our actual range we will see more Axx flops too

We will see these boards more often and so we can bluff more often which will help us get more folds than we would if we had played the A2 in the first place.

If we choose to go for the pre-adjusted range, except this time the board *misses* our perceived range, we can maximise our value by showing up with a value hand the opponent doesn't expect

If the board comes low we'll sometimes reach showdown with a straight and will take value from even the best hands in the villain's range.

If the board comes TTx, 99x, 55x, 56x, or anything similar, will appear as though we have almost certainly missed but secretly we are quite likely to have hit.

And if we *hit* when the opponent thinks we are sure to have missed, we rinse them.

Pre-adjusting lets us to make profit on more boards, it gives us greater coverage in the future.

Not only do we make money when our perceived range hits, as is standard, we also make money when we appear to have missed.

Pre-adjusting gets even better! Because it's also a very sneaky play to make

It's not as easy to read as a normal bluff.

The opponent will find it very difficult to switch up his play to account for the cards that we've shown him, because, he will very rarely see what it is that we are doing.

He won't know that we folded our small Ax Pre-flop.

He'll only learn that we truly held 67 after losing a load of money to it and even then he will fry his own brain trying to work out what it is that we'd done.

When he looks to our range he'll see that we're playing with the percentage of hands that he expects us to play with. He'll rack his brain trying to work out how we turned up with cards that he didn't expect. He'll probably start thinking we're cheating well before noticing what we're really doing.

If he's unaware of how to pre-adjust, which most players are, then he will usually go completely insane. He'll scream abuse and call you a fool. "You should never have held those cards!"

His cognitive dissonance will make it almost impossible for him to work out the reasonably simple tactic which we have employed.

Even if he does eventually realise that we're making this play against him without those low Ax hands, and that we're replacing those hands with the unexpected suited connectors, if he's still incapable of consciously understanding *how* to pre-adjust himself then he will only be able to counter our play in this one type of situation.

We will be pre-adjusting in almost every situation

Almost every single time we make a decision with a future card still to come we benefit by folding with some of the weaker hands that will be in our perceived range and continuing instead with cards that the opponent thinks we would not possibly hold.

If he does recognise what it is that we are doing in any one situation; it's our job to notice, change our consideration of our perceived range, skim off some of the weakest hands in our re-evaluated perceived range and then replace those cards with others which he won't expect.

Pre-adjusting can seem complicated. It is *pretty much* the most complicated part of this decision-making method. But remember, you have been using this same method for your entire life

All you need to do is *recognise* how to pre-adjust.

Then, with very little practice, you'll suddenly realise that you're adept in its application. You'll be able to apply Level Three far more easily than you'll be able to explain it, that's for sure.

Pre-adjusting is hugely important in all games

We often see commercial businesses competing with one another to provide the same kinds of product. If you find yourself running a competitive business like this then you can torment your competition by applying pre-adjusted strategies.

Any time you introduce unexpected new products you are pre-adjusting.

Any time you put any plan into action behind closed doors, you're pre-adjusting.

To make the pre-adjustment as profitable as possible you need to make it as difficult as possible for the competition to adjust to contend with you once they realise what it is that you have done.

Step 10 - Consider how your perceived range will look during future decisions. This enables you to show up with cards that the opponent won't expect.

BET-SIZING

It's only once we get to Level Three that we find a reason to make a bet in Poker. If we didn't have a subconscious understanding of our perceived range we wouldn't be able to consider any aggressive play.

We need to be capable of recognising the hands that we're representing or we'd have no reason to make any bet.

The opponent doesn't fold for no reason. Subconsciously we know this. He folds to our bet because he thinks that our cards are probably stronger than his in relation to the price, or that they will be by the time the hand reaches conclusion.

If we're playing No-Limit Holdem we can devise our own bet-size

At Level Three we concoct this size based on all the factors that you have learned up until now.

Our perceived range, the villain's range, the potential profits etc etc.

Against a low-level villain, if we *don't* want him to continue in the pot with a hand, then we need to design our bet-size so that he has the *wrong* price to continue against the range that we ourselves are representing.

If we do want the low-level opponent to continue with a hand, then we design our bet-size so that he has the right price to continue against the range that we're representing.

For us to make an accurate assumption of the range that our bet-size is representing, our sizing must work in coordination with the situation that we're in –

- Imagine we were up against a player who is not consciously aware of his perceived range, Level Two or below, and we see a board which puts both a lot of draws and a lot of value in both of our ranges.

If we held a strong draw – the opponent will usually expect us to bet big so that we can make chips from folds whilst protecting our currently weak showdown hand.

This is a commonly learned play.

At the same time as making money from hands in the villain's range that will fold, we'd also be attempting to build a bigger pot so as to improve our implied profits once we do hit our hand.

If, instead, we held a strong value hand – he's likely to think that we'd bet just enough to make him feel as though he is getting the correct price to make a straightforward play and continue on with his draw.

We can then assume that if we're to bet around the correct price to make our opponent continue with a draw, we'll be representing a value hand.

If we bet bigger than this, we'll be representing a draw ourselves and the opponent might well continue with all kinds of weak hands.

- Now imagine we're in the opposite kind of situation, a situation where the board brought no big hands nor draws to either player. In this situation, we can't validate a very big bet. There are very few hands in our range which we would want to bet big. Even if we held the nuts, we wouldn't want to bet big because there would be very few hands in the opponent's range which we'd expect to call.

Most competent low-level players would therefore start off by believing us to be bluffing if we made an expensive play on this sort of board.

I'm certainly not saying that we shouldn't bet big on these dry boards against low level players, I'm only saying that we probably shouldn't do it while we're attempting to bluff against these particular players.

Bet-sizing is a highly un-linear factor in No-Limit Poker
We can be very creative with our sizing.
It can be quite fun to play with our opponent's understanding of our range using different bet-sizes, especially when we're against accurate Level Two players!
I often find times to make small bluffs against these guys to represent a slightly stronger hand than their own holding. I love to see them make what they think is a great fold to what they think is a good value play from a very specific range.
The Level Two players won't consciously be aware that we understand the cards that we are representing so we can toy with them forever and they may never learn how badly we take advantage of them.

What we're doing with our sizing is playing around with our perceived range to make our opponent do what we want him to do with his range so that we can make the most profit
It's important to keep those potential profits at the forefront of our deliberations.
Our goal here is to make profit from the hands in his range.
If we have a very strong hand and we think the villain will continue with a large part of his range when facing a big bet, this is usually the best size to choose. (To be perfectly accurate we would have to consider the future moves so that we could see our profit come showdown.)
If on the other hand, we are weak and can get him to fold lots of his range to a small bluff, then that is usually the best size to choose.
The chips need to be considered in the same way as ever, and so the smaller we make the bluff, the less of the villain's range we need to fold in order for it to be a profitable bet.

There is a lot to be said for standardising our bet-sizes too
If we don't change our bet-size depending on our holding then we don't give the opponent any clues as to the part of our range which we currently hold.

This does restrict the control we hold over the opponent, but it will also make him struggle to perceive an accurate range for us.

I certainly wouldn't use standardised bet-sizes against a beginner who wouldn't notice my sizing anyway.

Against this kind of player, I'd usually bet big with my strong hands and small with my weak hands.

It's also worth mentioning that it's of great benefit to make these beginners familiar with our larger than average bet-sizes so that they will assign us a broader perceived range whenever we make an expensive play. We certainly don't want them to be thinking we're strong whenever we bet big.

Step 11 - Design your bet-size according to all factors, whilst paying close attention to the primary goal - making more profit on average against the opponent's range.

THE STANDARD PLAY MODEL

And here is it! The standard way to consider a poker play. In this chapter we combine of all the factors that you have learnt up to now. We put it all together into one decision-making formula.

At the higher Levels this play model doesn't differ a great deal from the one which we have here at Level Three. If you're able to learn this, you will easily be able to adjust it to cater for any play at any Level.

This play model could also be adjusted so that it applies to *any* problem in *any game.*

The physical factors from Level One are interchangeable. You can replace the cards and the chips with other physical factors related to your own problem.

Only the rules are specific to each individual game, if they are changed then our relative physical attributes change along with them and then that will filter through to all the other Levels.

The formula itself always stays the same.

There are three main stages to the decision-making process:

1. First, we use the past to gather information so that we can find the present situation.
2. Then we imagine the future changes to the present situation after we have selected our possible manoeuvres.
3. Finally, we compare our options and make the manoeuvre that will make our present situation better.

In Poker:

1. We establish the ranges. We also consider the chips.
2. We consider what will happen with the ranges depending

on our possible plays. We also consider what will happen to the chips.
3.   We select the play which will make us the most chips.

Making a play at Level Three is very similar to making a play at Level Two
Our goal is still to make the most money on average against all the different hands in the villain's range.
At Level Three we need to add our perceived range to our deliberations. We also need to add to our calculation the money we make by getting the opponent to fold.
As Level Three gives us the *reasons* behind Level Two, we need to consider our own perceived range before considering the villains range.

In this coming example I'll start by giving you the full calculation which will enable you to give yourself an exact figure for your profit. I'll skip past this full sum as quickly as I can and then I'll give you a much easier method which you can use to calculate your plays at the tables.

Imagine we're pre-flop, and are wondering whether to call, raise, or fold to a low-Level Reg's re-raise:

Stage 1

(Level 3) ... My perceived range is 99+, AJ+
(Level 2) ... Villains range is JJ+, AQ+
(Level 1) ... I actually have KK.
He has raised me 15 chips and has 75 left. I have 90 chips. And there are 35 in the pot.

Stage 2

If I opt to re-re-raise all in...

The price of the play is 90, if he calls the pot will contain 200 chips.

My perceived range will become QQ+, AK.

So, the villains:

AK will call and I'll win around 70% of the time. From the 200 in the pot, I win 140. The play costs 90, so my profit is 50.

AQ will fold, my perceived range is too strong and the price too high. So, I will win 35.

AA will beat me almost every time so I lose - 90

QQ loses to me almost every time so I make 200 - 90 = 110

JJ is a little tricky to pinpoint. I don't know if he will fold or not as he is getting about the right price to call against my perceived range. Let's guess that I'll get around 70 in profit from this hand.

AK and AQ will appear 3.5 times more often than the pocket pairs, and so here is the final sum:

(AK) 50 x 3.5 = 175
(AQ) 35 x 3.5 = 122.5
175 + 122.5 - 90 + 110 + 70 = 387.5
387.5 divided by these 10 instances = average profit of close to **40**

If I opt to call…

The current price is 15, the pot will be 50.

My perceived range will be TT - JJ, AJs – AQ

… So, against;

AK – I will almost always win provided we don't see an Ace which will appear around 1 in 5 times. So, 1 in 5 times I lose 15. And then 4 out of the 5 times I will not only win the 50, I also win an extra 25 or so when he makes an inevitable bluff against my future perceived range. So, let's say we make 75 when we win. 75 x 4 = 300...... 300 - 15 = 285....... 285 / 5 instances = 60ish
The price of the play is 15 so on average, I will make around 45 against this hand with the call.

AQ - I win even more than I do with the AK because if he hits any of the Queen's in the future he is probably going to put in all his money. Those A high boards are still bad though. So, 1 in 5 times I lose 10... 3 in 5 times I win 75... 1 in 5 times I win 110.
3 x 75 = 225...... 225 – 10 + 110 = 325...... 325 / 5 instances = 65
The price of the play is 15 so on average, so we make around 50 from this hand.

AA - I lose, but maybe not a full stack. Let's guess my losses at around 70 on average against this hand.

Against QQ - I win about the same as against AA, but he could bluff me on those A high boards. So, let's say that I make around 50 from this hand.

Against JJ I win on average, but he could bluff me with it on those A high boards again. So again, let's say we make around 50 on average.

(AK) 45 x 3.5 = 158
(AQ) 50 x 3.5 = 175
158 + 175 - 70 + 50 + 50 = 433

We then divide this by all the instances, which again is 10, and we find that we will make somewhere around **43** chips profit from this call.

Stage 3

In this spot, we appear to make slightly more from the call on average, and so we call.

*****

This calculation is very difficult to do accurately at the tables. In this example, I even left out parts of the sum to make it easier for you to read -
We don't know if the opponent will bluff us all the time with his missed hand which would change our winnings.
I haven't accounted for the times the opponent hits his AK on the river and continues betting. Which would lessen our winnings with the call against AK.
I haven't accounted for the times the board brings that last King.
I haven't calculated for all kinds of small factors which would slightly change our winnings.
In this example, with this calculation, we should call, but this is a very close call and if we were to go into more detail regarding those future cards and future moves against any specific villain we may well find that the raise is the preferable play.

The way I would calculate all this at the tables is much easier

To find out whether or not we need to fold all we need to do is make sure that we will make some profit.

So, if we are considering a fold all we need to do is compare the price to the win chance, like we did at Level Two.

In this example, our price on the call is 2:1, and as we're beating way more than a third of the villains range we currently we have a good price to continue with our KK, and so we don't need to worry about folding.

As we're clearly in profit and are not folding, all we need to do is compare our profitable plays to see which is better:
AA beats us about the same regardless of whether we call or raise, and so we can forget about that hand.

AK gives us about the same profit too, so we can forget about that hand too.

The QQ and JJ in his range are important to us, because we think we can get more money from them if we raise.

But the AQ will give us more if we call.

We know that AQ comes up more often than QQ and so without three pages full of math we can already guess that both plays offer quite similar profits.

The math equations above are important to be *aware of*, but you don't need to do any of it accurately at the tables.

At the tables, all you need to do is make a rough guess as to how much you might win or lose against the hands in the villain's range.

You could calculate an exact profit number for each play, which is more accurate, but why bother! That sum is huge and our mind already has its own method.

All we do is weigh up our options.

We weigh up our win-chance against the price to see if we need to fold. We weigh up one profitable play against another to see which play is more valuable.

We just weigh up everything. The only other thing our mind needs is to add some weights. Some strengths, or some *values*.

In our minds, we don't need a number to tell us that one thing is more valuable than another.

In my mind, when I'm playing, I always picture the villains range and then it's almost as though I see his hands coloured either green or red depending on whether I expect to win or lose from it. The brighter the colour, the more I expect to win or lose from that hand.

As I trowel through all the different future cards and future moves I see those hands turning brighter or duller as I start to think they will be worth more or less.

All I need to do then is weigh up the red hands vs the green and if I can't get out of the red I fold. The more green I see the better I think the play will be.

Instead of calculating the numbers accurately, it's far more important to consider our opponent and all the future moves more accurately

In this example, if we were up against the kind of opponent who would always bluff his missed AQ and is likely to be gentle with his QQ when we saw an Ace high flop, the call will be the play that I chose.

If I had positional advantage and was up against a player who is likely to slow play the AA post-flop and thus save me a load of money then I would probably make the call to keep things cheap. If the player goes mad with his AK post-flop when he's missed, then I will probably call.

If the player has some chance of calling our raise with some low pair hands then I will raise.

If you're up against multiple opponents you consider everything in the exact same way

With more players in the pot, there are simply more ranges to consider, more chip stacks, more future actions, etc etc.

Whether you are in a casino, or playing online, whether you know lots about the opponent or a little, you always consider everything in exactly the same way.

To weigh up our options whenever we make a poker decision is surprisingly easy to do once you get your head around all the simple steps in this book

I do understand that chapters like this can be confusing, but to be a great player all you need to do is spend a few minutes making sense of those calculations and then the rest will come naturally to you.

After a little practice, you should feel confident following both your perceived range and the villains range to some degree of accuracy.

With a little more practice, you will feel comfortable pre-empting the changes to those ranges and designing plays that will, on average, make you money in the end.

The only thing that takes time to learn is accuracy, but accuracy is a mental attribute that we will be developing forever.

To be a Level Three player you need only be aware of these factors to *some degree of accuracy,* as soon as you can do that you will suddenly realise that you are comfortable choosing what to do when playing Poker.

For the most part, you will have learnt how to strategize in this game.

You will have completely unlocked Level Three in your mind, and should easily be able to make money from your poker play.

Provided, of course, you can keep your head.

Step 12 – Select your plays by weighing up your profitable plays against each other to see which will make you the most profit. If you can't find a profitable play, fold.

OUR PERCEIVED BODY/EMOTIONS

I have no qualms about pretending to the opponent that I, myself, am under mental distress. I do this all the time. It's easy enough to do, either online or at a casino.

If I've lost a big hand before hitting something strong I will often pretend that I am perturbed. I might start betting big, and fast.

People are far more likely to put money in the pot with a weaker hand if they think we're tilting. If we hit some good strong hands quickly after losing big we can quite often win back all our losses in mere minutes.

In live games, we have far more of an opportunity to deceive our opponent using our perceived emotions

You need to be somewhat of an actor but weaker players are certainly highly susceptible to this form of deception.

We might "Tut" again, or say something like "Oh man, what a load of *****".

We really can lead players into believing whatever it is that we want about our range, depending on how well we can act it out.

In live games, you have to love doing little fakeys too

I'll sometimes make it look like I've completely missed and then 'accidentally' go to throw my cards away while the opponent is still taking his move. If the opponent notices, as planned, I might well see his face light up as he bluffs away the last of his chips, followed by the disappointed bafflement when I call and show him strength.

This type of fake is effective, especially if we're able to gauge the opponent's reaction to it. If he sees that you're giving up but doesn't get excited and instead looks disappointed we might well presume that he has a strong hand instead.

We can do other fakeys too.
Another one I like is making it look like I'm about to put in a load of chips in the pot when it isn't my turn.
The opponent then thinks that I'm bound to make a bet and so he checks to me and then I sheepishly check it back.

We can toy with the opponents understanding of our plan in all kinds of ways using our body language/perceived emotions, but again, I do still feel it's important to remind you that we're entertaining these guys
We don't want them to walk away demoralized. We want them to feel exhilarated so that they come back to our table time and time again.
We don't benefit from humiliating them once our ruse is successful. We do benefit from being a gentleman, or a lady. We shake their hand or offer them commiseration with a smooth, confident smile.
We're already taking their money, it's only fair to let them leave with their dignity.

It is fun to mess with the opponents understanding of your emotions/body language in live games, but if you don't feel confident taking advantage of them in this way you could use an old trick to prevent the opponent making any physical read on you
You could do what they call 'the pot stare'.
If we simply stare at the pot while playing, or concentrate on making very few movements or signals what so ever, we would stop him reading anything regarding our range from our physical actions.
Unfortunately, if we do adopt the pot stare or something similar we would lose some of our control over the opponent. We also miss out on all the fun!

LEVEL FOUR

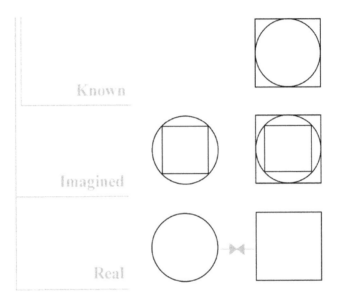

Relates to our perception of the opponent's perception of themselves from our perspective.
(Before I came along that was about the only description of Level Four you could find. It's hardly a wonder so few have cracked this Level)

Level Four is the last Level that gives us an advantage over our imperfect opponent
At Level Four we realise that the opponent has the ability to pretend to hold different cards to that which his actions lead us to believe.

We realise that he could be bluffing.

We realise that he could have pre-adjusted his range.

The way we consider this new development is by considering a new range for the opponent. A range consisting of the cards he is pretending to have.

If we believe the opponent is meddling with his perceived range, we can consider that which he is pretending to have alongside our own perceived range to make a more accurate assumption about what he's actually holding:

- If he's trying to make himself look as though he has a strong hand while we look weak, the chances are that he holds something weak and is attempting to make us fold.

- If he's trying to make himself look weak while we look weak ourselves, we'd expect that he's attempting to make value off his strong cards.

Very few opponents are consciously aware of their perceived attributes and so we don't often need to go into much detail regarding Level Four while sat at the Poker tables, but, people can naturally bluff, and so Level Four is always a factor

Before considering what the opponent really holds, we can now consider both our perceived range *and* what it is the opponent is pretending to hold.

The range which we would have previously assigned the opponent at Level Two doesn't change; it's just that he has now become aware of it.

He knows what it looks like he's doing.

This *known* range no longer represents what we think the villain really holds, instead it represents what the villain's pretending to hold.

We then consider for a second time what it might be that the villain really has.

Is he pretending to be strong, or is he truly strong?

Has he pre-adjusted, or does he really hold what it appears?

This is easier to understand when matched up to the diagram
The first of the distorted squares, our standard perception of the opponent, has moved up a place and has become known to both players.
This range becomes the villain's pretend range.
It is then replaced with the new shape, our new presumption of the villain's actual range which has extra layers of distortions brought about by the extra range that is required for its creation.

These new developments don't change our game plan a great deal
The opponent may still be playing his cards in a straightforward manner if he thinks it's the play that will make him the most money. He might be making a straight forward play.
And we can still use Level Three in the exact same way as before.

Now that we're aware that the villain is capable of bluffing we become capable of inducing his bluffs which adds yet another dimension to the profits we can make in the game
We can now lure the opponent into bluffing us by making our range look weak which will make us extra value.
Or we can lure him into pretending to be weak with his strong hands which will help us keep the pot small in a situation where we expect we could lose more.

Even though it appears there are additional chips to be made once we can consider Level Four, there are actually less
We have found more ways in which we can make money from the opponent, but we have also realised that the opponent is capable of a higher Level of thought.
At each Level of the opponent's development we lose potential profit.
Gradually, as the opponent becomes more proficient at the Levels, we lose more and more of our advantage over him.

Another thing that holds back our profit is the fact that the

game has become more complicated
I could write about individual Level Four situations forever but there would be little point.

Once you become capable of considering the additional range, the villains pretend range, you'll be able to explore all the possibilities for yourself and will be able to create your own counters to your opponent's deception in real time whilst you play.

I will, however, explain to you how Level Four logic effects our most simplistic hand selection. This will give you a better grasp of the changes that will occur once you become capable of calculating this penultimate level:

- At Level Two we're playing hands that are stronger than our opponents.

- At Level Three we split up our actual range adding hands which the opponent is unaware of.

- At Level Four we need to create an extra split in our range. We need to play the cards which take value from the cards which the villain is playing that he expects we are unaware of.

If the opponent is slyly playing 56s – 9Ts, then we might add hands like 55 and 66 to our actual range. Perhaps TJ or TQ.

It's completely up to you and depends on the range the opponent is using, what he's like, how many chips he has, what he might do in the future etc, etc.

Everything depends on everything.

You might be interested to hear that Level Four is nowhere near as profound as Levels Two and Three
Even if you're an experienced player, when you come to recognise this Level you won't notice that enlightened feeling as intensely as you did for the other Levels.

We don't come to realise anything as profound as empathy or

awareness here, and so Level Four isn't nearly as exciting as the lower levels.

Having said that, when we recognise Level Four we do acquire one *significant* skill.

For the first time we're able to see the pattern that exists between the levels.

Once we reach Level Four, for the first time, we become capable of *understanding all the Levels*.

We become capable of understanding all Poker

It was only once I'd recognized this Level that I was able to separate the first three into their proper groups.

I hadn't pinpointed Level One prior to recognizing Level Four.

Level Four holds all the keys to understanding everything regarding Poker, strategy and the minds calculative process.

Once you become aware of this Level for your specific game you can see for yourself how the Levels co-exist and this makes you capable of breaking down any play at any Level.

Once you reach this Level, you're fully ready to play against anybody. All the Levels become unlocked and you'll have no problem consciously accessing the minds strategic formula whenever you like for your game.

Strangely, at this point, we suddenly begin to see the Levels *all around us*.

Which really is rather surreal…

When I read through all the statements in Sun Tzu's famous strategy book 'The Art of War' I was instantly able to match up each piece of strategic advice with its corresponding Level

After completing this task I actually re-read the entire book more carefully as I felt that it'd been way too easy.

But sure enough, after the second time through I came to the exact same conclusion – Sun Tzu, the *greatest* strategist of all time, only recognized as high as Level Three.

To be more precise, he reached a Level Three understanding of war and then he resorted to Level Infinity. He missed out Level Four.

THE LEVELS

If you've heard of Sun Tzu you will probably think I'm being arrogant. I thought so myself!
And so I went over all the evidence again and again.
I discussed it with others.
I searched outside the box for a solution thinking that perhaps something had been lost in the translation.
Yet still I found nothing to contend with my first assumption.
I could only conclude one thing – Sun Tzu was not consciously aware of Level Four.
It's quite clear to see.
He doesn't mention any factor in his book that relates to Level Four, which isn't damning evidence, but he also made a clear mistake. A mistake that a Level Four player would never make.
It's almost painful to read.
He appears to have been suffering from cognitive dissonance when he wrote it out. He sounds troubled, (*perhaps* due to losing troops in the past). He sounds unhappy with his own advice.
Even a beginner can see the flaws in his logic yet he himself was unable to comprehend it.
See for yourself.
The mistake I'm speaking of is in this statement:

"If the enemy leaves a door open, you must rush in."

Do you think we could destroy Sun Tzu's entire army by leaving a door open that led off a cliff!?
The answer is obviously no.
The reason why the answer is no is because Sun Tzu would subconsciously be capable of Level Four.
He would quickly realise that we had deceived him, he would see his mistake in this situation and would not force his army to jump to their death.
I do wonder though, how many times I could leave different doors open before he realised that his logic was flawed.
Any situation where he faced a door would allow us an advantage over this man in a battle, but there are many more situations

possible in war than are possible in Poker.

For each individual situation we could apply Level Three logic or above against Sun Tzu and I expect that we would eventually obliterate him.

Just like we would at the poker tables.

All warfare is based on deception… Hmmm… Sounds to me like Sun Tzu had too much pride.

Step 13 - Consider the Villains pretend range. The cards that he is pretending to hold.

LEVEL FIVE

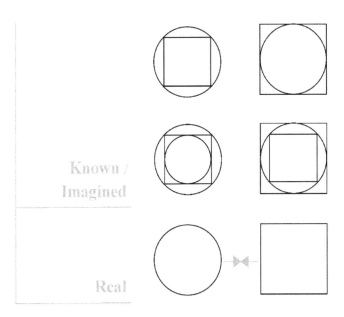

Relates to our perception of the opponent's perception of our perception of ourselves

Now that both we and the opponent are capable of understanding all the different factors relating to both of our perceptions of each other, we finally become capable of knowing one another.
There is no longer any advantage to be gained from applying a higher level of thought than the opponent's aware of because we're both aware of *all levels of thought*.
The deception we once relied on suddenly dies.

We both now know what we are both capable of.
When two excellent players square off against each other anything can happen. The whole game opens up. There is literally so much that each person is capable of doing!

Level Five adds to our deliberations the standard changes that we come to expect from a higher level
We create an extra range for consideration.
Our own pretend range.
We also gain an extra level of bluffing capability. We now know that the opponent is aware of our perceived range, so we can now pretend to bluff rather than actually bluff.
(If we're against an opponent who is competent at Level Four he will be completely aware that we're capable of this and so the advantages are minimal).
Once we reach Level Five the thing that becomes most valuable is accuracy. Not only in our mathematical equations but also:

- How accurately we understand the opponent,
- How accurately we understand the opponent's perception of us,
- And how accurately we understand the opponents understanding of our perception of him.

Every single factor I've described in this book becomes a platform on which we utilise all our strengths or weaknesses as a person to gain or lose accuracy
If we have a keen attention to detail then we'll find it easy to notice subtle differences in the opponent's plays which will help us break down his range.
If we are creative we'll be able to design more effective lines to maximise our profit against his range.
If we're arrogant we'll often be misled.
All players need to deal with emotions on a day to day basis and so wisdom becomes very important by the time we reach this Level.

THE LEVELS

The ability to drive the opponent insane becomes one of the largest factors that gives us an advantage over him.

No matter how skilled he is, no matter how well he knows us, if he's frustrated then he will not be able to calculate all that he must in order to defeat us.

At Level Five the strategy we use to select our plays changes for the last time

For the *first* time, we find reason to play the *same hands* in the *same situations* but in *different ways*.

We now benefit from adopting a mixed strategy with our range.

Sometimes we play our hand one way, other times we benefit from playing it a different way.

Our opponent understands all our moves and so we mix up our plays to make it as difficult as possible for him to know what we hold.

In regard to our basic hand selection, we now need to add an extra split in our actual range

We do this in the same way as we'd come to expect from the other Levels except now we would also mix-up our strategy with individual hands.

Sometimes we'd raise our AA, sometimes we'd call with it.

By the time we're capable of this Level we realise that we, or the opponent, could be playing any reasonable hand in any way at any time.

It's no longer a matter of whether or not we think the opponent would have chosen to play a certain hand a certain way, but rather the *chance* that he might have chosen to play it that way

The likelihood that he would hold certain hands depends mostly on the strength of those hands.

He would obviously still be certain to play AA pre-flop for the value come showdown, but now we don't know which *way* he will choose to play it.

Now there is always some tiny chance that he'd choose to play 72 as well. Of course, ordinarily, it is extremely unlikely that the opponent would ever select to play such terrible cards. If the situation was ideal and the field of opponents perfect he probably still wouldn't play them often if at all.

The thing is - By the time we're aware of Level Five we don't really need to worry about making mistakes anymore, because we know that the opponent could well hold anything at any given time.

We don't just play random cards. The other levels *still* haven't changed. We're *still* trying to beat his range using our cards, our perceived range, etc etc.

It's just that there is no longer such a thing as a correct way to make a specific play anymore.

By the time we are fully aware of Level Five we will have learnt exactly how to play Poker in its entirety and so from then on we are free to just play.

In actual fact, the thing I find most profound about this Level has very little to do with Poker

I presume it's at this Level that the mind becomes capable of completing its original task -

At Level Five we truly become capable of understanding any object in its entirety

That doesn't mean that if you can consider an object at Level Five then you completely understand it. Rather, it means that once you're able to consider any object from a Level Five perspective you become *capable* of understanding it.

All factors relating to all objects exist within levels Two to Five

Level Six does still exist but nothing changes. There are no new factors to consider aside from those which we already expect from a higher Level.

I can only guess as to the real profundity of this information. Perhaps the people out there who are creating artificial intelligence will find it fundamental. Perhaps psychologists will someday realise that it's revolutionary.

As for me, I'm simply aware that this is where the importance of the Levels *ends*.

The entire purpose of creating these Levels was so that we could understand the effects of our interactions with an unknown object. At Level Five we truly become capable of it.

Step 14 - Consider our own pretend range. The cards the opponent thinks we're pretending to have.

THE FAMOUS CONTINUATION BET

I think it'll be nice to end the exploitative levels with a play, and there is no play more important than the Cbet

(Now that we've reached Level Five we can add all our bluffs and the villain's bluffs to our deliberations. In this coming play, I could add a description of the range that the opponent thinks it looks like he holds, his Level Four range, and we could add our Level Five range too, but that would be very complex and would make my lovely page look all messy, so instead, I will choose an opponent who isn't consciously aware of these factors so that they won't be much of an issue.)

Let's imagine that we're sat on a 6-seated table and are the first to act pre-flop when we receive;

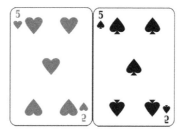

We open betting by putting in 4 big blinds, and then we're called by the button -

The player on the button is very tight, he always plays at these same $100 cash tables, he's not very creative, he seems to play with a set strategy pre-flop and mostly for value once the flop comes down.
(This is easily enough information to tell us that we are most

probably facing a weak Level Two opponent - He doesn't change his plays in the early stages of the game, which is weak, and he doesn't seem to know how to do anything but shoot for value, which is Level Two. This guy is too dull to be aware of Level Three. Any bluffs he makes will be memorized plays which we can find and exploit.)

And then the board comes;

Our perceived range: A5s+ (ATo+), KTs+ (KJo+), 22+ - 67s+
Villains range: ATs+ (AQo+), TJs + (KQo+), 22 - JJ
Chips: 96 each, 11 in the pot.

If we bet around 1/2 pot -

Our perceived range: KTs+, 33/QQ/KK/AA, Bluffs like AT/TJ

Our perceived range has smashed this flop. We've hit it with loads of our hands.
If we now opt to bet this villain will usually think that our range is strong. If we hold the K or better, we would be attempting to take value from his Q, his weak K or his draw, and we would want to protect our hand against his draws.
He might also think that we're bluffing with the odd draw ourselves.

Villains range:

His ATs and AJs will most probably fold to the bet, and we take

the pot.
All those missed pocket pairs are almost certain to fold to the bet, and we take the pot.
TJ will continue, and I think we will most probably lose to it in the future. He will probably bluff us off our hand even if he misses his draw.
KQ will continue also, and obviously there is the rare 33 too.
His AK, which he might well have raised preflop, and his AQ are an annoying part of his range. They will probably just call against our perceived range and then eventually beat us.

I can't see many reasons for us to put any more money in the pot on any future card other than the 5, which only comes one in ten times, and so implied profits don't look much different from our current profits.

To summarize - It looks like he is going to fold to our bet with almost half of his range, and as the price of the play is 2:1 we are making money!

If we don't bet;
Our perceived range may contain a Queen, but mostly it looks like we've completely missed and are scared that he has hit. We're not attempting to protect our medium strength hand. We're not attempting to take value with a strong hand. We look like we have something like 55.
His entire range will destroy our 55 with all his future plays. If we check, even this guy will realize that he can bluff us if he needs to.

Solution is simple. We bet.

*****

The Continuation bet is among the first plays that any professional player learns

If we're the last player to raise up the price before the flop, when the flop does come down we will be the only player at the table whose standard perceived range contains all those AA type monster hands. If we are to then *continue* betting, it scares many an opponent into folding more than they should.

The Cbet is a very common play, and so many opponents do have ways to play back against it, but not many know how to play back well.

If they don't understand all the poker theory that I have described then they are sure to make loads of mistakes.

Most players don't read boards properly - they don't consciously consider how often any particular range hits.

They don't change their plays against at individual players.

They don't follow the ranges properly.

Etc, etc, etc…

You're going to walk all over all of them.

Usually your opponent will have simply memorized specific situations where they can play back against your Cbets. They will memorize boards which they can attack. They will memorise specific plays.

If you can figure out which plays he has memorized for which situations you can destroy him even more efficiently than usual by using this exact same decision-making method.

Sometimes, in this kind of spot, I do check and show the opponent I hold a weak hand, but only because I feel really sorry for the poor guys!

This kind of player will lose against us all day long and that must feel terrible. It's only nice to let them out play us for the odd little pot.

If we do check here and then give up, this guy will be happy and for good reason too!... He will also be overconfident. A check and a then a fold here will help us hide in the shadows.

LEVEL INFINITY

The dark Level.

In 1950 a man called John Nash witnessed a girl being attacked outside his dorm-room window when he discovered a mathematical solution known as 'The Nash Equilibrium'.

What he had found was a strategy for games which is unbeatable. He had found Level Infinity.

We poker players were quick adopt his strategy and over the decades we've gradually refined our understanding of it.

Nowadays it's known in the poker world as GTO. Game Theory Optimal.

Which sounds awesome, right...

To calculate a play at this Level, to calculate a GTO play, we need to follow a range which I haven't *really* highlighted yet. Our actual range. All the hands that we could actually hold in any given situation.

Unsurprisingly, every single one of us is already familiar with this new range. We use it all the time.

Anytime we consider any general strategy. Anytime we think, "I will play more aggressively", we are designing a strategy using our actual range. Anytime we design a strategy against any specific player we concoct it using our actual range. Any work we do off the table analysing players to creating strategies is all completed using our actual range.

All strategies are designed with our actual range, and GTO is just another strategy.

Except this one's different, John Nash came up with a strategy

that could not be beaten
Using exploitative poker theory as I have described it, you can see what he worked out.

He *basically* realised that as the Levels advance the actual hands selected for the plays become more and more balanced between strong hands and weak hands.

He then noticed that the natural end point is where each player would have as close to an equal amount of strength and weakness in their actual range as is possible in any given situation.

What Nash had realised was that if we hold an equal amount of strength to weakness at any given time there is literally nothing the opponent can do to beat us.

Imagine it like this.

Imagine the opponent tells us exactly what his actual range is going to be.

Yet instead of telling us that he will have a strong range or a weak range he tells us that he will have exactly the same amount of strong hands as he does weak regardless of what we do.

There becomes nothing we *can* do.

On average, we can't bluff him with any benefit to ourselves; we can't play for value with any benefit to ourselves.

The only thing that we can do is exactly the same as that which he is doing. We are forced to balance our own range.

This wouldn't be the end of the world if GTO were as good as it sounds, but it is not
The way it works is by sucking all the profit out of the exploitative Levels.

If it's us that is applying this strategy our opponent can gain no advantage by applying Level Two, which sounds amazing, but, there is a MASSIVE drawback: If we are applying this strategy then we ourselves can't gain any profit from Level Three!

We throw all our deception out the window.

We lose *all* our control over the opponent.

If either myself or my opponent is applying Level Infinity then all the potential profit from all the other Levels disappears for both of us.

The only way any profit could be made is when one player or the other diverts from equilibrium and even then, the profit margins are minimal.

GTO is only supposed to be used a defensive strategy, which is only supposed to be employed when there is no opportunity to exploit, but even in this they've got it wrong -

There is always an opportunity to exploit. No human is capable of perfecting a GTO strategy in No-Limit Holdem. There are simply too many situations out there and too many factors to take into consideration which means that we can always exploit anyone.

There is absolutely no point in ever attempting to play using an equilibrium strategy, instead, we should spend our time learning to exploit better instead.

If we use the exploitative theory as I have described in this book against a player who is balancing, we would find ourselves balancing too.

We wouldn't actually need to learn how to balance. It would just happen naturally.

Both roads lead to Infinity, but if you take the exploitative study route as I have described it in this book you will develop your mind and maximise your own profit.

With no care for my personal opinion, almost every professional player from my generation learns GTO

And most of them have absolutely no idea what they're doing.

You see, to make a GTO play in this day and age you don't actually need to know how to make a GTO play.

If we *did* want to calculate our GTO play we would have to consider our actual range and balance the strength, weakness and draws in accordance with the price we have for continuing.

Not only that, but we'd also have to consider all the future changes which are yet to come and anticipate the future strength of all our possible hands so that our range would continue to be balanced as best as possible regardless of what might happen.

There is no simple way to calculate all of this for every single situation.

It's so difficult to calculate, that the Nash solution to No Limit Holdem has only recently been discovered.

Well, technically, it's not true to say that they've found the Nash solution to the most popular version of Poker, but by using an obscure equation they seem to have gotten so close that they might as well have done.

Since then, computer programs have been developed which allow the users to compare their plays with the GTO play.

These programs enable players to memorise as many GTO plays as they like without learning the first thing about Poker theory.

As the player memorises more and more Level Infinity solutions to different situations they slowly learn to balance.

If you're fully aware of the exploitative Levels I can see no reason why memorising GTO plays would have any unpleasant side effects

But if you're not fully aware of each Level, memorising these plays will quickly make it more and more difficult for you to approach your missing pieces of theory.

Not only would you have knowledge of plays blocking your understanding of the Levels, you would have perfectly *incorrect* knowledge blocking you. A GTO play is the one play that we would *never* want to make. And so if we build trust in Level Infinity plays before understanding the other Levels it is catastrophic for the mind.

Cognitive dissonance goes into turbo mode!

If you speak to a Level Two player who has spent many years memorising unbeatable GTO plays, you won't be able to mention Level Three factors without them ripping off your head and spewing down your throat.

They themselves are missing out on *all* the profit from Level Three but if you try to teach them they just go crazy.

These guys *can't even comprehend* the control they're forsaking by memorising thousands upon thousands of unbeatable Level

Infinity plays.

It really is awful.

It wouldn't be so bad if the problem stopped with them, but by applying GTO when it's clearly not needed they are minimizing the potential profit on the table as a whole.

Not only are they themselves missing out on a shedload of cash, all the players at the table are missing out on cash.

Even the casino misses out.

The entire Poker market has its feet swept from under it.

These crazed players don't stop at verbal abuse

They constantly attempt to re-assure themselves that GTO is superior by convincing other players that it is better than any other strategy, and, as the exploitative Levels give us the only other strategies, these guys desperately attempt to dissuade others from exploiting without even realising what they are doing.

I swear it's like invasion of the body snatchers out there!

Everyone is singing the praises of GTO to everyone, beginners are applying it before they even know what GTO stands for. And if you dare to mention anything about exploitation everybody turns on you.

They even call GTO plays "correct" plays!! It's madness.

In the very lowest stakes games the players are attempting to play with a balanced preflop range nowadays because they've learnt from some book somewhere that this strategy is correct.

None of these books explain that it will massively decrease the amount of profit that the player themselves stands to make.

And the reason they don't explain it is because most of the authors are all mad on GTO too!

It's literally everywhere.

These GTO guys have even changed the definition of a bluff so that it works for their deception-less style. Which is something I plan to amend as soon as possible.

There are a couple of reasons you might want to learn GTO

I've already mentioned that if the opponent is using GTO we need to use it ourselves or we lose, and so there is obviously some benefit to understanding what an equilibrium strategy looks like. It's useful to know as much about the opponent's strategy as you can before sitting at the table against them.

The *only* other reason we might learn this GTO strategy is so that we can avoid it like the plague.

Many of the GTO players themselves know that their style is sucking their own profit from their profession but they can comprehend no other way to make their plays and so continue on

GTO is a very hard habit to break.

And it is not only the poker players who are using it.

Game theory is huge nowadays and equilibrium strategies are a fundamental aspect of it.

Stock brokers, economists, businessmen, strategists of all kinds. All these rather important people use game theory nowadays and I would bet a pretty penny that many of them are deliberately spewing our society away.

If these guys are learning Level Infinity strategies before understanding the other Levels then I expect they will play at Level Infinity needlessly even if they know full well that it is killing their own profession.

I suspect that GTO a huge problem in the financial world of today.

Level Infinity is like *The Anti-Game.*

It relentlessly consumes the life blood of games, spreading from player to player eating up all the potential profit until there is none left.

It rapidly decays everything it touches, and by teaming up with cognitive dissonance it even corrupts the players to increase its own growth rate.

Fending off John Nash's beautiful mind and educating the world regarding the true effects of his equilibrium strategy has been among my primary goals over these last few years

I would have loved to discuss my logic with the great

mathematician himself. He had always been plagued by mental illness and I felt that there was a chance I could have brought him some peace by showing him some solutions to the problems that his equation had brought about.

But, unfortunately, the demons that had hold of him had a different plan.

On the very same day when I finalised the diagram which carries this book the remarkable John Nash and his wife tragically died in a car accident.

I'm aware of three things that we can do when an unbeatable strategy has dominated over our game

1) When GTO took over online chess, the players stopped playing and played something else.

When it took over backgammon, the players stopped playing and played something else.

When you play tic-tac-toe, (noughts and crosses), equilibrium is quickly realised by both players and it isn't long before every game ends in a draw. And so what do we do? We stop playing. We play something else.

The most obvious way of overcoming this equilibrium is to stop playing the game. Pull out our assets and move them over to a different game.

2) The second way is to teach players how to exploit. This will widen the gap between a beginner and GTO which will improve the games potential for profit and thus extend its lifespan.

3) Or, finally, we can introduce new rules. Such as; banning GTO.

Now that GTO has been deciphered it shouldn't be too hard for a casino to enforce this rule and it would instantly make the unbeatable strategy obsolete

We will all be able to continue playing the game we love without programmed players ruining it for all of us.

I expect that the implementation of this new rule will trigger the next major advancement in the professional world of Poker.

As soon as it is introduced a new generation of Poker player will

find their home.
GTO will be a thing of the past. And we will *all* make loads of profit.

When I think about the financial world, I can't come up with any reason to continue using money (trade)
We should just stop playing this silly game, we should pull out our assets and play something else.
We should find some other way to reward each other for our efforts.
I'm not talking about communism or socialism, I'm talking about capitalism but without the money. Otherwise known as paradise.
Instead of a system based on debt where everyone is owed for their efforts we can easily flip it all around and make a system based on generosity.
A simple website that records everyone's contributions to society as well as their requests from society could well replace trade all together.
The more and more I think about it, the better and better this idea seems, and I've had it on my mind for many a year now.
On the website we promote generosity by rating everyone's contributions to us, and if a contributor trades, their overall rating goes down.
Simple. The site will need a little fine tuning, but the basic logic seems to work far more effectively than our current system.
I might make this website myself, but that does seem to be beside the point. I'm not a web developer and in this new world it would be a web developer who contributes the website.
If you currently make websites for a living then please do move forward with this plan. Once your sites up and running I will put up my contribution and I'm sure it will earn me many a reward.

Final Step – Learn strategies and plays. From books, from coaches, and most

importantly, from your opponents. Provided you are fully aware of the other Levels, learning GTO plays will not hinder you.

... I don't think it'll hinder you. But then, who could fully understand these Levels anyway!? I think I'm gonna' stick with them...

ONE LAST THEORY

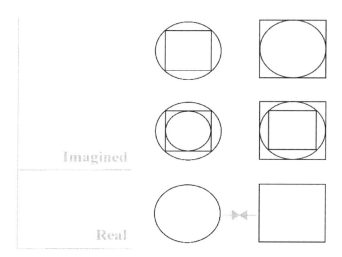

Buddha said that all physical things are manifested by the mind.
If we're going so far as to say that this formula holds an unrecognised yet fundamental place in the manifestation of all physical things, it wouldn't be much more of a stretch to say that by analysing the formula we could develop the way we understand subjects like physics, and mathematics.
If we're able to make more sense of these two subjects by using this formula then perhaps we can add credibility to the lessons of a legend.
You never know, this formula could hold the key to understanding everything. We might be able to use it to help us understand how and why we are alive...

As we advance through the Levels our mind develops new abilities-

- At Level One we learn about our strength and our capabilities.
- At Level Two we learn about our opponent and how to defend against him.
- At Level Three we learn about our image and how to attack.
- At Level Four we learn how the Levels work, and how to utilise this information to maximise the benefits from our interactions.
- At Level Five we become an equal with the opponent.

All of these advancements have an interesting goal in common. They all serve to help us understand Level One.
Our self.
A new secret regarding our self is revealed at each new Level, and by the end, we learn what to do with ourselves.

The second I noticed Level Infinity I was surprised to see that it held one *last* Level One secret
A strange secret.
This is not completely unknown to the poker world. Many professionals have already heard of it and they don't think it particularly profound.
To me though, it made *too much* sense…
Level Infinity, seems to be, completely identical to Level One
If a person were to play Level Two perfectly they'd need to be capable of playing all the other Levels perfectly as well. If a person was to play Level One perfectly they'd not need any of the other Levels, they would simply be playing at Level Infinity.
Infinity is the perfection of One…
One and Infinity are the same…
And that got me thinking. What if the object, our opponent, was infinitely good?
What if the opponent had all possible attributes and performed every manoeuvre perfectly.

What if our opponent were some kind of God.
This is obviously the best opponent we could pit ourselves against but to see the *full* potential of the minds formula we would also need to consider ourselves starting off as nothing.
So…
Nothing vs Infinity.
0 vs 1 perhaps???

When I fed these entities into that diagram it seemed as though I could see a story emerge
A story of how a non-existent entity becomes an infinite being.
From a grain of dust to a living thinking person.
All so that someday we would become an equal to some kind of Infinite 'thing'?
When I realised that those two numbers were the first in the Fibonacci Sequence the story started making *way too much* sense.
The Fibonacci sequence is a numerical pattern which currently befuddles scientists.
Perhaps you've heard of the golden ratio, which is basically the same thing.
The entire universe was made with this golden ratio in mind but nobody knows why this has happened.

The number sequence goes like this…
0 , 1 / 1 , 2 , 3 , 5 , 8 , 13 … with no end.
The line / that I added shows the divide between the real and the imagined if we were to feed this Fibonacci number sequence into the minds formula.
That which is on the right would be the mind's way of making sense of that which is on the left. The numbers on the right would be an imagined reflection of those on left.
The number 0 would relate to nothing.
The number 1 would relate to everything, or infinity.
The number 2 would relate to two infinities, as the mind divides infinity into bitesize chunks.
And the number that we normally call infinity would be the

imagined reflection of 0. The imagined version of this non-existent entity.

When I laid the newly filled in diagram over Pascal's Triangle, another fundamental numerical pattern, the story made enough sense for me to completely abandon this line of study before I went completely insane
All of it was based on a guess that was vague at the best of times. I'm no mathematician.
I don't know if the number One could be identical to Infinity.
I don't know if the other numbers could all be imagined.
All I have to go on is this voice in my head telling me that all of this makes sense alongside everything else I have come to believe is most likely true.
If nothing more, crazy theories like this show us just how far the applications of this formula might stretch.

The mind is everywhere, it's in everything we see and everything that we do
Maybe the real goal for life is to become an equal to some kind of infinite being.
Perhaps all the trust we're putting in science is actually *blocking* us from recognising our true cognitive capabilities.
Rules that we've come to know, like gravity and time, could be *restricting* our true potential.
Maybe, if we quiet our mind, we will realise that cause and effect don't work the way we think they do. That time is not omnipresent.
We might then notice that all existence is imagined.
That there is no difference between ourselves and every other object.
That we already are at one with everything and just haven't realised it yet!
Maybe all the emotions we experience every day are caused by cognitive dissonance because in truth, our God-like mind already understands all the things that we cannot yet

comprehend
Maybe all the bibles and all those prophets got it right.
Maybe we're all the same Adam, sitting under a tree, wondering why we're perfect.
Maybe Buddha *really did* crack it all.
Maybe he did become an equal to a God... or maybe he didn't.
We mere mortals can but guess.
Everything effects everything else, and so the only way we can be certain of anything is if we are certain of everything.

In this book I've attempted to describe all that I have learned from a simple mind game
By now I suspect that you realise I am no psychologist, nor am I an economist, a mathematician or a theologist. I'm not even much of a writer!
I'm just a poker player.
I hope this book won't get slated until it's stomped into the mud.
I hope that one among you specialises in another field and can take this *formula* further.
This thought-process was weirdly difficult to decipher and so I do hope everyone makes the best use of it.
As for me, well, I have all kinds of plans. Chief among them is taking over the poker world, and I do think I stand a good chance of pulling it off.
With GTO infesting everything related to Poker the vast majority of companies in this industry are choosing to duck away from what they think is a dying market.
I can hardly believe my luck.
All my competitors are stepping aside to leave the entire market ripe for plucking.
There is literally too much for me to do by myself.
The poker market needs new forums, new guides, new casino games, new software. New everything, which all follows one new rule - GTO is banned. (Or at least contained).
My next move is to create "The Original Playbook". If nothing changes by the time that is complete, I will make my own forum

too. And eventually I plan to open my own casino.

I don't believe this game is on its way out
It looks to me like it's just getting started.
If we do replace money in the years to come I expect games like Poker will rise from the dirt.
They will be seen in a completely different way.
The professionals wouldn't be known as reckless gamblers.
Players would play for pride, to have fun and to learn.
With this in mind, it even seems possible that games like Poker could eventually be revered.
Maybe one day we'll see monks wandering the land carrying only a deck of cards and a sack of chips as they challenge one another to battles of the mind.
That would be something.
That would be change.
That would be progress.
And that's what we all want.

ABOUT THE AUTHOR...

Growing up with three brothers and six sisters is as hectic as you'd imagine. Without a TV to drain our attention we would play endless games. Scores of score sheets would flutter around the house as we continuously wrote lists of points, winners and losers. We would play and play. Winning, losing, learning.
Carl was the eldest of us and so was the keeper of the pride. I was the third child, and as the second son I felt it my duty to challenge my elder brother at every given opportunity.
And we certainly did not lack in opportunities. We lived in a house where you had to fight for a space on the breakfast table. We were never savage, we didn't maul each other. We were all very well-disciplined and well composed kids, but we were always at war.
In our youth, Carl and I battled our way up to black belts in Tae Kwon Do, and if we two ever met in a competition it was always epic. I'd almost always lose, but any time I landed a blow we both knew it was worth more than one of his. He was top of his class at school, had three years on me and half a foot of reach, so I didn't stand much of a chance at pretty much anything.
Over the years I took my fair share of the wins and so did everyone else. My army of sisters certainly didn't sit quietly in the shadows. Perdi, the eldest of the sisters, was generally my ally back in those days but if we did ever find ourselves facing one another it would usually be me who walked away with my tail between my legs. The next sister down the line is away in Alaska racing huskies as I write this now. She's a blackbelt too. Right this second she's probably wrestling a polar bear to save some baby seals.
Anyway, my elder brother always loved strategy games, and he was always a big favourite to win.

He left our hometown when he was 18 to live with our father. This is about the time when online Poker arrived in cyberspace and it wasn't long before he was playing it ferociously. As was I, of course.

Living in different countries we were both learning the game in very different ways. He was studying training guides and working with international coaches while I played constantly and learned from my opponents as I taught myself.

One day my brother called me and explained that he was living in Bulgaria playing and teaching Poker. He said that he was making a killing and I wanted in.

I left my job and went to see him.

Although I didn't understand it at the time, after a few months in a professional environment I was playing at Level Two. It took me around a year to reach Level Three and by then I was making an average of £50 per hour.

Money was something I'd never had and now it was rolling in. I moved back home to England and started living the *successful* lifestyle.

I reached Level Four as my daughter and I moved into the flat of our dreams. Flash and swanky.

But then something unexpected happened.

Exposing my opponent's weaknesses had always been a skill that I was proud of, but something was different now. As soon as I noticed a player's weakness I found myself feeling really bad for them! Taking their money suddenly seemed wrong! It didn't seem fair.

The weakest players literally didn't stand a chance against us and so I began to *feel sorry* for them.

Losing at the poker tables isn't fun, and now I wasn't enjoying winning either!! Reaching the highest skill level had made me so confident in my ability that I ceased to enjoy making my living from playing anymore.

I had always been competitive and ruthless but now I kept finding myself rooting for my opponents. I wasn't in the right frame of mind to play, I had lost my drive, and so started to back away from the tables.

That is about the time when I stumbled across Buddhism. Through Buddhas lessons I came to believe that I would be better off teaching Poker rather than playing.

I took on a couple of practice students, I taught/discussed theory on online forums and I also continued teaching my elder brother. When I had reached Level Four, Carl was still stuck at Level Two. I was confident that I understood my game and I explained it to him very clearly and very often. He was desperate to match my win-rate but no matter how I approached it he always refused to accept that Level Three made any sense.

He was a clever man. Well versed in Poker. Strategy games were his forte and he had spent a long time learning this particular strategy game, but his mind blankly refused to accept anything above Level Two.

He would even get particularly irate with me anytime I mentioned anything to do with the higher levels.

This seemed strange, but it wasn't that that *really* baffled me.

It took me two long years to teach my experienced brother all the Levels.

But I'm able to teach a beginner in a matter of minutes.

I hope with this book that I have managed to teach you.

Yadi, 2017

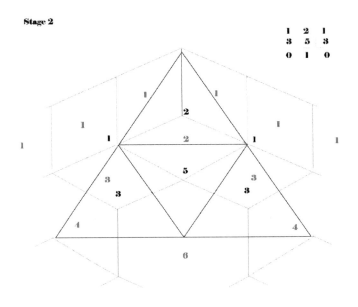

Stage 2

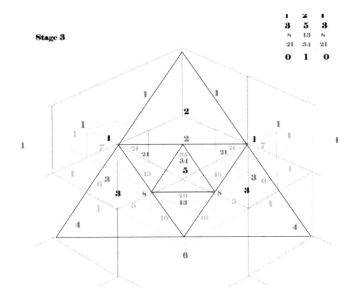

Stage 3

I'd like to give a special thanks to my editor Mitch.
Mitchelwicking@gmail.com
He asked for nothing to work on this book. He is a man who truly loves his work. What a legend!

A huge thanks to my family too. For helping me in a hundred different ways whilst I completed this extremely difficult project.

Lightning Source UK Ltd.
Milton Keynes UK
UKHW05f1913130218
317812UK00009B/130/P